W9-BEK-064

What Einstein Didn't Know

What Einstein Didn't Know

Scientific Answers to Everyday Questions

ROBERT L. WOLKE

A BIRCH LANE PRESS BOOK
Published by Carol Publishing Group

Illustrated by Diana J. Zourelias

A Birch Lane Press Book
Published by Carol Publishing Group

Birch Lane Press is a registered trademark of Carol Communications, Inc.

Editorial, sales and distribution, rights and permisisons inquiries should be addressed to Carol Publishing Group, 120 Enterprise Avenue, Secaucus, N.J. 07094

In Canada: Canadian Manda Group, One Atlantic Avenue, Suite 105, Toronto, Ontario M6K 3E7

Carol Publishing Group books may be purchased at special discounts for sales promotion, fund-raising, or educational purposes. Special editions can be created to specifications. For details, contact: Special Sales Department, Carol Publishing Group, 120 Enterprise Avenue, Secaucus, N.J. 07094.

MANUFACTURED IN THE UNITED STATES OF AMERICA

10 9 8 7 6 5 4 3 2 1

LIBRARY OF CONGRESS CATALOGING-IN-PUBLICATION DATA

Wolke, Robert L.
What Einstein didn't know : scientific answers to everyday questions / Robert L. Wolke.
p. cm.
"A Birch Lane Press book."
ISBN 1-55972-398-X
1. Science—Miscellanea. I. Title.
Q173.W787 1996
500—dc20 95-48082
CIP

To my personal energizers: daughter Leslie, who has energized my teaching by continually asking "Why, Daddy?" and wife Marlene, who continually energizes my life by being a true partner.

Contents

Introduction

Forget the word *science*. This book simply tells you what's going on beneath the surface of everyday things. It is for people who are curious about the world around them, but who don't have the time to seek out the explanations or may be a bit intimidated by *science*.

Of course, the answers to why everyday things happen must be scientific—that is, logical and accurate. But you won't find the usual pop-science nonanswers here that leave you just as mystified as before. Instead of mere *answers*, you will be given *explanations*: plain talk that I hope will lead you all the way to genuine, well-I'll-be-darned *understanding*.

Traditionally, people have encountered science in four places: classrooms, textbooks, children's books, and in deadly serious tomes by solemn scientists. Unfortunately, science classrooms and textbooks have turned off at least as many people as they have turned on. (Don't get me started.) The fun books for kids are great, but they promote the false notion that only kids can be curious about things. And the solemn science books only perpetuate the conviction that science is inherently incomprehensible to ordinary mortals.

This is not a textbook, it is far from solemn, and it is not a fun book for kids. (But don't be surprised if your kids steal it from you.) It is a fun book for grown-ups. But it is not a collection of gee-whiz facts to be marveled at and instantly forgotten. Instead, it answers real questions that might be asked by real people in real situations: around the house, in the kitchen, in the garage, in the marketplace, and in the great outdoors.

There is no need to read this book in sequence. Browse to your heart's content and peruse any question that catches your eye; every explanation is self-contained. But whenever some closely related information exists elsewhere in the book, you will be referred to the question and answer unit in which it is explained.

As you browse, you'll see a number of Try Its, tests and demonstrations that you can do yourself, whether seated at your kitchen table or on an airplane. You will also find a number of Bar Bets that may or may not win you a round of drinks, but will certainly spark a lively discussion.

Throughout the book, whenever an explanation threatens to become more than you want to know, the details are banished to a Nitpicker's Corner for easy skipping. Occasional technical buzzwords are explained when they are used. But if you should stumble upon one and forget what it means, you will probably find it in the list of Buzzwords at the back of the book.

Okay, so you've looked ahead and seen the word *molecule* popping up almost everywhere, and you're afraid that the explanations might be too technical. Fear not. *Molecule* is just about the only technical buzzword that is absolutely unavoidable in explaining your everyday surroundings. You may already have a pretty good idea of what a molecule is, but for the purposes of this book, here is all you need to know:

• A molecule is one of those invisibly tiny particles that everything is made of. All the things that you see and touch are different because their molecules are of different kinds, sizes, shapes, and arrangements.

• Molecules are made of clusters of even tinier particles called atoms. There are about a hundred different kinds of atoms, and they can combine in an enormous variety of ways to form an enormous number of kinds of molecules.

As Keats might have put it, "That is all ye know on earth and all ye need to know, going in."

Happy understandings.

As a professor, I have concluded many a class by asking, "Are there any other questions?" But this time I'm asking you, my readers. If you have any questions about everyday science that you'd like me to answer in future editions of this book, just send them to Scientific Answers, 610 Olympia Road, Pittsburgh, PA 15211. Or, you can e-mail them to *wolke+@pitt.edu*. If your question is used in a future edition, I'll give you credit by name and city, so be sure to include that information.

What Einstein Didn't Know

1

Around The House

Let's just wander about the house for awhile. If we have our antennas up, we'll find a wealth of fascinating things to look into. Perhaps candles are burning on the dining table and the champagne is bubbling away while we admire a sunset through the picture window. Or maybe we're stuck down in the laundry, where soap and bleach are working their chemical magic on all that ooky stuff that we lump under the general heading of "dirt."

In this section, we'll see what astounding things are going on in the candles, the champagne, the sunset, the soap, and the bleach, not to mention the water bed and the shower.

Here's the Dirt on Soap

They say there are three things you don't want to see being made: sausage, laws, and soap. I've already heard too much about legislators, and I'd rather not know about sausage, but I'll brace myself: How do they make soap?

The unholy mess involved in making soap belies its use as an incomparable cleaner of just about everything for at least the past two thousand years. It has always been easy to make out of cheap, readily available materials: fat and wood ashes. Lime was sometimes used also.

You can make it the way the Romans did: Heat limestone

to make lime. Sprinkle wet lime onto hot wood ashes and mix well. Shovel the resulting gray sludge into a caldron of hot water and boil it up with chunks of goat fat for several hours. When a thick layer of dirty brown curd forms on the surface and hardens upon cooling, cut it into cakes. That's your soap.

Or, if you prefer, just go to the store and buy a cake of today's highly purified commercial product. In addition to soap, which is a definite chemical compound, it probably contains fillers, dyes, perfumes, deodorants, antibacterial agents, various creams and lotions, and lots of advertising. Sometimes more advertising than soap.

Every soap is made by the reaction of a fat with an *alkali*—a strong, potent base. (A base is the opposite of an acid.) Instead of goat fat, today's soaps are made from any of a number of different fats, including beef and lamb tallow and the oils of the palm, cottonseed, and olive. (Castile soap is made from olive oil.) The alkali used in making today's soap is usually lye (caustic soda, or sodium hydroxide). Lime is another handy alkali, while wood ashes can still be used in a pinch because they contain the alkali potassium carbonate.

Having been created by the addition of an organic compound (a fatty acid) to an inorganic compound (lye), the soap molecule retains some features of both its parents (see p. 63). It has an organic end that likes to fraternize with oily organic substances, and an inorganic end that is attracted to water (see p. 98). Hence its incomparable ability to coax oily dirt into the wash water.

Whenever you see the following chemicals listed as ingredients on the label of a shampoo, toothpaste, shaving cream, or cosmetic, be neither alarmed nor impressed; they're all just the chemical names of soaps: sodium stearate, sodium oleate, sodium palmitate, sodium myristate, sodium laurate, sodium tallowate, and sodium cocoate. If the "sodium" is replaced by "potassium," the soap has been made with caustic potash (potassium hydroxide) instead of with caustic soda (lye, or sodium hydroxide). Potassium soaps are softer and may even be liquids.

Cleanliness Is Next to . . . Impossible

Whenever there is something on our bodies, clothes, or cars that we don't like, we say they are "dirty" and we wash them. What we call dirt can be any kind of stuff at all. But soap always seems to oblige us by removing it, and only it. How does the soap "know" what's dirt?

It would appear that soap is a magic substance that recognizes and respects our skins and precious possessions while devouring everything else under the sun like a vulture leaving only bones behind. But no such magic substance exists. Instead, the answer has to do with the natures of oil and water. Simplistic as it may sound, everything that we call dirt—more politely, "foreign substance"—is either oily or is stuck to us with oil. And soap (see p. 3) is a uniquely good oil remover.

Before we can figure out how to remove dirt, we must look at how we get dirty in the first place.

A microscopic speck of dirt—meaning anything we don't want attached to us—can be stuck on in one of two ways: either it is mechanically trapped in a microscopic crevice or else it is moist, and the moisture makes it adhere. An example of the former is the kind of dirty you get on a dusty road; an example of the latter is the kind you get on a muddy one. In either case, a good hosing down with plain water, encouraged perhaps by a little rubbing, will do a reasonable job of removing the foreign substances. Soap isn't really necessary.

But what if the dirt particles have a slightly oily coating instead of a watery one? They will stick to your skin just as the wet mud did. In fact, the dirt doesn't even have to bring along its own oily coating. There is often enough oil on our skins to make dirt particles stick. Unlike the mud, however, this dirt is going to stay stuck, because oil doesn't evaporate and dry up as water does. Nor will a spray of plain water dislodge it, because water won't have anything to do with oil (see p. 98); it will simply roll off the dirt as it would off a duck's back, which, as you know, is covered with oily feathers.

It seems, then, that the only thing we can do to unstick oil-adhering dirt is to seek and destroy the sticky oil itself. The dirt will then be able to fall off or be swept away by a liquid.

Well then, let's fill the old bathtub with alcohol, kerosene, or gasoline; they're all good solvents for oil, aren't they? That's what dry cleaners do to our dirty clothes: They tumble them around in a barrel full of a solvent such as *perchlorethylene,* or *perc* for short, an organic solvent that is a phenomenal dissolver of oil. They call the process "dry" cleaning in spite of the fact that it involves sloshing things around in a very wet liquid. The thinking seems to be that if it isn't water, it isn't wet. Wrong, of course (see p. 194).

Unfortunately, perc in the bathtub would kill you even faster than the alcohol, kerosene, or gasoline would, so we can forget about bathing in solvents. But there is one substance that is just as good, and it's not very toxic (mouths have reportedly been washed out with it): soap. Soap doesn't actually dissolve oil. It accomplishes the astounding feat of enticing the oil into the water, so that the oil and its captive specks of dirt can then be flushed away.

Soap molecules are long and stringy. For almost all of their length (the "tail") they are exactly the same as oil molecules and therefore have an affinity for other oil molecules. But at one end (the "head") they have a pair of electrically charged atoms that just love to associate with water molecules, and this head is what drags the whole soap molecule into the water—makes it dissolve. While swimming around in the water, if a gang of dissolved soap molecules encounters an oily particle of dirt, their oil-loving tails will latch onto the oil, while their water-loving heads are still firmly anchored in the water. The result is that the oil is pulled into the water; its captive dirt particle is released from whatever it was stuck to and can be swept down the drain.

NITPICKER'S CORNER:

There is a second important thing that soap does: It makes water wetter. That is, it helps the water to penetrate into all the nooks and crannies of whatever it is we're washing.

Water molecules stick to each other quite strongly (see p. 98). As a result, a water molecule situated at the surface of a "piece" of water has very strong attractions that are trying to pull it into the rest of the "piece." Now, the tightest huddling-together formation that any group of particles can achieve is to gather into a spherical shape; a sphere has the smallest possible amount of surface exposed to the outside world. That's why water forms spherical drops whenever it is free to do so, such as when it is falling as rain.

(In two dimensions, that's why the pioneers "circled the wagons" against the Indians; if they had "squared the wagons," they would have been exposing more of themselves to the outside.)

This inward-pulling force on the surface molecules of a liquid is called *surface tension*. It arises because the surface molecules are, in a way, different from the molecules in the body of the liquid.

In the body of a liquid, a molecule is pulled upon by attractions to fellow molecules above, below, and all around it, and these pulls cancel each other out. But a molecule right on the surface is pulled upon only from below and all around, but not from above; so there is a net downward pull, uncancelled by any upward pull. This makes the surface molecules adhere more tightly to the water than the other molecules do, and the water behaves as if it had a taut skin on its surface. Small objects can even lie on the surface without sinking through the "skin." And water bugs can even skate merrily along on the water's surface.

Enter soap. Soap molecules disrupt the surface tension of water by crowding around the water surface with their water-loving heads in the water and their oil-loving tails sticking out. This disrupts the huddling-together tendency of the water molecules and allows them to pay some attention to—that is, adhere to and wet—other things . . . including a floating needle.

TRY IT Because of surface tension, you can rest a steel sewing needle on the surface of water in a bowl. Lower it down carefully with a couple of toothpicks or match sticks.

After you've gotten the needle to rest on the water's surface, sprinkle some powdered laundry detergent near it, but don't actually bomb it. Detergents are even better than soap at killing surface tension. As soon as some of the detergent dissolves, the needle will plunge precipitously to Davy Jones's locker.

Love-Boat Logic

A TV commercial for a Caribbean cruise line says, "For our guests who may have been out in the sun too long, we even wash our sheets in soft water." Is that for real?

No, the advertising copywriter has been out in the sun too long. Makes you wonder whether a cruise line that swallows this kind of foolishness from its advertising agency is capable of finding the right islands.

Rather than insulting my readers by pointing out why the sheets will not be any softer, I might gently remind any prospective cruisers among you that hard and soft water are not called that because of their relative rigidities. Nor is it because you make hard-boiled eggs with one and soft-boiled eggs with the other. The choice of "hard" and "soft" as sobriquets for these waters was unfortunate; they would better have been called "difficult" and "cooperative"—with respect to soap.

Hard water is water that has been around the block. It first fell through the air as rain and then frolicked and percolated over, around, and through the rocks and rills before being apprehended, detained, and exploited by humans. In its peregrinations, it inevitably picked up carbon dioxide from the air, which made it acid: carbonic acid.

This acid can dissolve small amounts of calcium- and magnesium-bearing rocks such as limestone (calcium carbonate) and dolomite (mixed calcium and magnesium carbonates). It can also dissolve certain iron-bearing minerals slightly. As a result, the water can wind up containing dissolved minerals such as calcium, magnesium, and iron.

It is considered "hard" because it is hard for soap to do its job properly in water that contains these minerals. Soap consists of long molecules that have an oil-loving end and a water-loving end (see p. 5). It does its cleaning job primarily by joining oil and water together.

The trouble is that calcium, magnesium, and iron react with

the water-loving ends of the molecules to form insoluble, white waxy curds that effectively remove the soap from the water and prevent it from doing its job. These curds are evident as "soap scum" or, in their most infamous guise, "bathtub ring." (Contrary to folklore, the latter is more a sign of the hardness of the water than of the hygienic habits of the bather.)

Incidentally, here is our candidate for the most disquieting thought of the week: *You have probably eaten soap scum in certain candies*. One common form of soap scum goes by the chemical name of *magnesium stearate*; the stearate part comes from the soap (see p. 3), while the magnesium part comes from the hard water. Magnesium stearate is a soft, smooth, waxy substance. That's what makes it stick to the bathtub, all right, but that's also why it imparts a creamy texture to soft mints and other "suckable" candies—a soapy texture, if we may be so bold. If you see magnesium stearate listed among the ingredients of your candy, be assured that it is the pure chemical compound, manufactured from sources quite different from bathtub scrapings.

But back to the hard water. We can do two things about soap's inability to perform its job well in hard water: We can soften the water or we can lose the soap and use a synthetic detergent instead.

Water-softening tactics are based upon removing the offending minerals or rendering them ineffective. Many home water-softening units remove the minerals by *ion exhange*. Ion exchangers replace the calcium et al. with sodium, which is innocuous because it is already a part of the soap molecule.

In what seems like ancient times, about fifty years ago, hard water was combatted by adding washing soda (sodium carbonate) to the laundry tub. This chemical re-forms the original, insoluble calcium and magnesium carbonates—essentially, the original rock—thereby removing them before they can gum up the soap.

These days, however, practically nobody uses soap for laundry. The acres of washday products on the supermarket shelves are all synthetic detergents (and are all essentially identical, except for the hype). Like soap, they have oil-loving and water-loving ends on their molecules, but they simply refuse to react with calcium and magnesium. For good measure, they usually contain water-softening chemicals such as phosphates and—guess what—washing soda.

Hard water is still a villain, however, because it can clog up water pipes and boilers. When hard water is boiled, the dissolved calcium and magnesium fall back out of the water as limestone and dolomite. This born-again rock—called boiler scale—can form a tenacious coating on the insides of boilers, water heaters, and pipes, clogging them up like the arteries of a Viennese pastry chef.

If your water supply is hard, shine a flashlight inside your dry tea kettle and you'll see the boiler scale as a white coating on the surface. If it bothers you, boil some vinegar—an acid—in the kettle to dissolve it.

TRY IT Shake up a few shavings of bar soap or a few flakes of Ivory Snow (which is real soap) with some distilled water in a jar. You'll get a beautiful, thick head of suds, indicating that the soap is doing its job. (Distilled water is pure and free of minerals; you can find it in many supermarkets and drugstores.)

Next, if you live in a hard-water area, add some tap water and shake again. (If your water supply is soft, you can simulate hard water by adding a little milk instead.) The calcium in the hard water (or milk) will kill the suds flat. You may even see some soap scum in the form of floating white curds.

Burning a Candle at One End

When a candle burns, where does the wax go?

Except for what drips all over your tablecloth, it goes to the same place that gasoline and oil go when they burn: into the air. But in a chemically altered form.

Candles are usually made of paraffin, which is a mixture of hydrocarbons, substances that we find in petroleum. As the name implies, hydrocarbon molecules contain nothing but hydrogen atoms and carbon atoms. When they burn, they react with oxygen in the air. The carbon and oxygen become carbon dioxide, while the hydrogen and oxygen become water. (But not necessarily completely; see p. 13.) Both of these products are gases at the temperature of the flame, and they just go off into the air.

We burn many other hydrocarbons: methane in natural gas, propane in gas grills and torches, butane in cigarette lighters, kerosene in lamps, and gasoline in cars. They all burn to form carbon dioxide and water vapor, and seem to disappear in the process. Paper, wood, and coal contain additional mineral and plant materials that don't burn, so besides producing carbon dioxide and water they leave an ash.

NITPICKER'S CORNER:

When there isn't quite enough oxygen available to make a full complement of carbon *di*oxide, as in an automobile engine, we get some carbon *mon*oxide also; (see p. 106).

If you find it hard to believe that flames produce water, try this:

TRY IT Put some ice cubes in a small, thin aluminum saucepan, let it get cold, and hold it just over the flame of a candle or a butane cigarette lighter. After a while, check the bottom of the pan and you'll see that water vapor from the flame has condensed there into liquid water.

You didn't ask, but . . .

Why won't a candle burn without a wick?

By capillary attraction, the wick leads melted wax up to where it can be vaporized and mix with oxygen in the air. A

block of solid wax, or even a puddle of melted wax, won't burn because the wax molecules can't come in contact with enough oxygen molecules; only as vapors can they mix intimately, molecule for molecule, and react. Combustion (burning) is a reaction that releases heat energy. Once it begins, it gives off more than enough heat to keep melting and vaporizing more wax to keep the process going.

Fire!

The flames in my gas grill are blue, but the candles on the dinner table burn with a yellow flame. What makes flames different colors?

It's a matter of how much oxygen is available to the burning fuel. Lots of oxygen makes blue flames, while a limited amount of oxygen makes yellow ones. Let's look at the yellow flame first.

A candle is really a very complex flame-producing machine. First, some of the wax must melt, then the liquid wax must be carried up the wick, then it must be vaporized to a gas, and only then can it burn—react with the oxygen in the air to form carbon dioxide and water vapor (see p. 11). This is far from an efficient process.

If the burning were 100 percent efficient, the wax would be transformed completely into invisible carbon dioxide and water. But the flame can't get all the oxygen it needs to do that just by taking it out of the air in its immediate vicinity. The air, with its flame-nourishing cargo of oxygen, just can't flow in fast enough to take care of all the melted and vaporized paraffin that is ready to burn.

So, under the influence of the heat, some of the unburnable paraffin breaks down into tiny particles of carbon, among other things. These particles are heated by the flame and become luminous; they glow with a bright yellow light. And that's what makes the flame yellow. By the time the glowing carbon particles reach the top of the flame, almost all of them have found enough oxygen to burn themselves out.

The same thing happens in kerosene lamps, paper fires, camp fires, forest fires, and house fires: yellow flames, all. Air just can't flow in fast enough to make the fuels burn completely to carbon dioxide and water.

TRY IT If you don't believe that there are tiny particles of unburned carbon in a candle flame, just insert the blade of a table knife in the flame for a few seconds, to catch them before they burn out. The blade will acquire a deep, velvety black coating of carbon. This carbon black is just about the blackest substance known, and is used in inks.

Gas grills and gas ranges, on the other hand, start out with a gaseous fuel—no vaporizing required. That makes it easy for the fuel to mix with lots of air, so that the burning reaction can go full blast. Because the fuel is burning almost completely, we get a much hotter flame. And it's a clear, transparent flame because no glowing carbon particles clutter it up.

Want hotter yet? Why not mix pure oxygen, instead of air, with the fuel gas? After all, air is only about 20 percent oxygen. Glassblowers use a torch that mixes oxygen with natural gas (methane), to produce a flame temperature of about 3000 degrees Fahrenheit (1600 degrees Celsius). A welder's oxyacetylene (oxygen plus acetylene gas) torch can reach about 6000 degrees Fahrenheit (3300 degrees Celsius). Clear, blue flames, all—except when the torch is improperly adjusted so that the gas doesn't get enough oxygen to burn completely. Result? A yellow, sooty flame.

You didn't ask, but . . .

Why should a hot, properly adjusted gas flame be blue instead of some other color?

It has to do with the fact that atoms and molecules that are heated in flames can absorb some of the heat energy and then promptly spit it back out as light energy (see p. 168).

Every substance has its own typical wavelengths or colors of light that it emits after being stimulated by the heat. (Techspeak: every substance has its own unique *emission spec-*

trum.) The propane or natural gas in your gas grill and the acetylene in the welder's torch are very similar; they are all hydrocarbons—compounds of carbon and hydrogen. It happens that hydrocarbon molecules emit many of their particular light wavelengths in the blue and green parts of the visible spectrum. Other kinds of atoms and molecules, if they were vaporized and burned, would impart their own particular colors to the flame. That's how colored fireworks are made (see p. 168).

A Whoosh Is No Big Whoop

I usually buy my soda pop in two-liter bottles. But with such a big bottle, keeping the leftovers alive and fizzy from one pizza to the next is a problem. Besides keeping the cap on, what else can I do to keep it from going flat? What about that gadget that you put on the bottle and pump up? Does it really work?

Your objective is to keep as much carbon dioxide gas in the bottle as possible, because that's what the fizzy bubbles are made of. Keeping the bottle tightly stoppered certainly has to be your first line of defense. But frankly, it won't help very much.

There are many kinds of stoppers on the market, including that fancy pump-up job that you mention. It's a miniature bicycle pump that you screw onto the bottle, and then you pump a plunger to compress the gas inside the bottle. Sounds good but, unfortunately, it's a complete fraud. All it does is make you think your soda is livelier than it is. Let's see why.

Soda fizzes when dissolved carbon dioxide gas emerges as bubbles. The gas wants desperately to escape from the liquid because the folks down at the bottling plant have pumped in much more carbon dioxide than would ordinarily dissolve under atmospheric conditions. As soon as you open the bottle, most of that excess gas escapes into the room, and there is absolutely nothing you can do about that. Your only problem is how to make the remaining gas stay in the liquid for as long as possible.

Three things determine how much of a gas can remain dissolved in a liquid: the chemical reactions of the specific gas, the pressure, and the temperature:

- ***Reactions:*** Gases that react chemically with water will generally dissolve more readily than inactive gases, whose molecules have nothing to do but cruise aimlessly around in the water. Carbon dioxide is one of those gases that reacts. It forms carbonic acid, which adds that nice little pungent taste to soda, beer, and sparkling wine. Air (nitrogen and oxygen) doesn't react with water. As a result, carbon dioxide at room temperature is more than fifty times more soluble in water than nitrogen is, and more than twenty-five times more soluble than oxygen.
- ***Pressure:*** The effect of pressure is just what you'd expect: the higher the gas pressure above the liquid, the more gas will be pressed into the liquid. The way it works is that at higher pressures there are more gas molecules flitting about per cubic inch in the space above the liquid, and more of them will therefore be diving each second into the liquid.
- ***Temperature:*** The effect of temperature is probably just the opposite of what you'd expect: the higher the temperature, the *less* gas will dissolve. Saying it the other way, the colder a liquid is, the more gas it can hold. The reason for this is a little more involved than we want to get into right now, so we'll save it for later (See the "You didn't ask, but . . ." section on page 19.). But one example: At room temperature, water can hold only about half as much carbon dioxide as it can at refrigerator temperature.

Our conclusions, then, are that in order to keep as much carbon dioxide dissolved in the soda as possible, we must keep the gas pressure high and the temperature low. Temperature is no problem; we'll just make sure it's good and cold before we open the bottle, and then we'll put the leftovers back in the refrigerator as soon as possible.

But pressure is quite another matter. At the bottling plant, the carbon dioxide molecules were forced into the soda like a crowd of claustrophobes into an elevator. The instant we open the bottle, a frantic exodus takes place, and virtually all the carbon dioxide pressure goes off in one big *whoosh*! Once that happens, your soda is inevitably going to flatten; it's just a matter of time.

But is there really nothing we can do about that? Can't we restore the pressure somehow, that we may live to belch another day?

Enter the gadget hucksters. Just screw their gizmo onto the bottle, they say, pump the piston a few times, and there you are. Next time you open the bottle, you'll be treated to the biggest, most satisfying *whoosh*! you ever heard. And you're supposed to think that your soda is factory fresh.

But guess what? There isn't one more molecule of carbon dioxide in there than if you had simply screwed the cap on tight. You'd get the same big *whoosh*! if there were nothing but plain water and air in the bottle. The gizmo is nothing but an expensive stopper.

What you've pumped into the bottle is *air*, not carbon dioxide. Sure, there's a little carbon dioxide in the air, but it's only about one out of every three thousand molecules. The escape of a gas from a liquid can be decreased only by putting more of *that particular gas* into the space above the liquid. The amount of carbon dioxide that will stay dissolved in the soda depends only on how many collisions take place between *carbon dioxide molecules* and the surface of the liquid. If you had pumped in carbon dioxide gas, that would be another story; but nitrogen and oxygen are simply irrelevant.

Bottom line: Keep it capped and keep it cold. It's especially important to keep the bottle tightly sealed while it is out of the refrigerator, because that's mainly when the carbon dioxide is emerging, due to the higher temperature. So pour what you want, cap the bottle, and put it right back in the fridge.

But don't get your hopes up too high. You can slow down the exodus of carbon dioxide, but you can't stop it.

And oh, yes. Whatever you do, *never* shake the bottle. That only speeds up the emergence of the gas (see p. 31).

You didn't ask, but . . .

Why does warm beer go flat?

A larger amount of gas can dissolve in a liquid when it is cold than when it is hot. Or as a chemist would say, the solubility of a gas in a liquid increases with decreasing temperature. (But that's how chemists talk.)

In practical terms, why does the carbon dioxide choose to leave the beer just because it is getting warmer? From everyday experience, you might expect that as the liquid gets warmer, it should be capable of dissolving more stuff, not less. You can dissolve more sugar in hot tea than in iced tea, can't you? Then why should gases be any different?

The answer lies in the role that heat plays in the dissolving process. It can be a very complicated one.

When a substance dissolves in water, its molecules separate from one another and disperse themselves throughout the water. Other changes may occur at the same time, depending on the substance that is dissolving. For example, the molecules might attach themselves to tight little clusters of water molecules, or they may react chemically with the water, or they may split into electrically charged fragments, or do other things too horrible to contemplate.

All of these processes either use up or give off energy in the form of heat. So heat plays an intimate—and widely varying—role in the dissolving of various substances. The net result is that some substances will eagerly absorb the extra heat in hot water and use it to dissolve more, while some substances will react negatively to the extra heat and dissolve less. In other words, some substances will be more soluble in hot water than in cold, and others will be less soluble in hot water than in cold. Even chemists can't always predict which way it will go for a given substance.

In the case of gases, though, we can make a sweeping generalization: When gases dissolve in water, they *all* give off energy in the form of heat. You could say (and I will) that dissolving gases don't like heat; they're trying to get rid of it. So they will dissolve more readily in a cold, heat-absorbing environment like cold water, and they are discouraged from dissolving in a hot, heat-rich environment like hot water.

TRY IT Let a glass of cold water stand around for a few hours and, as it warms up, you'll see bubbles of air forming on the walls of the glass. The air was dissolved in the cold water, but the warmer water can't retain that much air. It "goes flat," just like beer.

Iconoclastic Elastic

Everybody knows that things expand when they're heated. But somebody wanted to bet me that there is a common substance around the house that will contract when heated. Should I have taken the bet?

No, you were right to pass on that one. The common substance is rubber. Stretched rubber.

Most things expand when heated for a simple reason: The higher temperature makes the atoms or molecules move faster (see p. 236). They then need more elbow room, spreading farther apart on the average, and that makes the whole substance take up more space.

But rubber can behave differently because of its oddly

shaped molecules. They're like worms in a can—thin, squiggly shaped chains, tangled all together into a disorderly snarl. That is, until you stretch the rubber. When you stretch it, the chains are stretched out straighter, forced to line up along the direction of the stretch.

But that is a very strained, unnatural state for them; you know that because you had to work to stretch them out that way, just as you would to stretch out a spring. As soon as you let go, the rubber molecules return to their compact, crinkled forms, and the rubber as a whole will snap back to its original shape.

What has that got to do with the effects of heat? Well, if you heat the rubber while it is in its stretched-out-molecule form, the heat-induced agitation of the molecules makes them pull in on their ends, which tends to decrease their length. (A wriggling snake is a shorter snake.) The rubber thus tries to revert as much as it can to its more compact form; it contracts.

TRY IT Cut a wide rubber band—at least a quarter-inch wide—to make a strip, rather than a loop. Use a tan, rather than a colored, rubber band; the colored ones generally aren't natural rubber. Tie a weight to one end of the strip and tack the other end to the edge of a shelf, letting the weight hang down freely. The weight should be heavy enough to stretch the rubber moderately. Now heat the rubber band with a hair dryer. Watch carefully and you'll see the rubber contract, pulling the weight up a bit higher.

BAR BET Rubber can contract when heated. That's stretched rubber, remember. Rubber that isn't stretched will expand when heated, just like anything else (see p. 37).

Beating the Heat

How can the very same thermos bottle keep hot things hot and cold things cold, seemingly at our whim? Someone told me it's done with mirrors.

To solve the problem, all you have to do is think of heat as a kind of liquid that flows only "downhill" from high temperatures to low temperatures. The thermos bottle acts like a dam that blocks the flow of heat. It won't let heat flow "down" from your hot coffee inside to the lower-temperature air outside. But by the same token, it won't let heat flow "down" from the outside air to your lower-temperature iced tea inside.

Another way of saying this is that the walls of the thermos bottle are a very effective heat *insulator*—a substance or arrangement of substances that retards the flow of heat. We're most familiar with using insulators to keep heat from flowing out of our warm bodies and houses into the cold outdoors; ski jackets, sleeping bags, and attic insulation come readily to mind. But our refrigerators are also insulated, in this case to keep heat from flowing *in*. Insulators work both ways.

Heat, of course, isn't a liquid, even though it does flow from one place to another. It moves in three ways: by *conduction*, by *convection*, and by *radiation*. Let's take them one by one and see how a thermos container foils them all.

Put a cool object in close contact with a warm one and you know what will happen: The warm object surrenders some of its heat to the cool one, so that the cool one becomes warmer and the warm one becomes cooler. Some heat has been transferred, or *conducted*, from the warmer object to the colder one.

But what is heat, anyway? It is the agitation, or movement, of an object's molecules (see p. 236). The more vigorously its molecules are moving, the warmer it is. So when you place a warm object (having rapidly moving molecules) in close contact with a cooler object (having slowly moving molecules), some of the faster molecules will collide with the slower molecules, transferring some of their energy to the slower molecules and speeding—warming—them up. That's conduction: direct molecule-to-molecule energy transfer.

When you touch a hot frying-pan handle, your skin molecules are speeded up by collisions with the frying pan's faster-moving molecules. When you touch an ice cube, your skin molecules lose some of their speed through collisions with the ice molecules.

A thermos container hinders conduction because it has double walls with nothing—a vacuum—in between. Because there are no molecules in a vacuum to collide with, heat conduction can't take place through it.

Convection is the process whereby heat is transferred from one place to another by the actual bulk movement of a heat-containing gas or liquid. You've heard people say that heat rises? What they really mean is that *hot air* rises, and along with it goes the heat it contains. That's convection. A convection oven is simply an oven with a fan in it that assists the circulation of hot air. In that case, the process is called *forced convection.*

A thermos bottle hinders convection simply by being a closed container; warm air can't pass through its walls. Any kind of closed container would stop convection.

Finally, heat can be *radiated* from one place to another in the form of infrared radiation (see p. 218). These energy waves are emitted by warm objects. They fly through space and can be absorbed by cooler objects, transferring their energy to them and heating them up.

A thermos container hinders infrared radiation by deflecting it with a mirror. The double walls of the container are silvered on their inner (vacuum-containing) surfaces, so any infrared radiation that tries to get through from either direction is reflected right back to where it came from.

If you think radiation isn't a serious contender for heat transmission, consider how you broil a steak *underneath* the heating element in an electric oven. Heat travels upward by convection, all right, but a lot of it also goes downward (and in all other directions) by radiation.

No thermos container is perfect, of course. Some heat is always being conducted or radiated out of your hot coffee or into your iced tea. But the thermos slows down the heat-transfer processes substantially, and your food or beverage stays hot or cold for hours, rather than minutes.

Incidentally, the name Thermos (it's just the Greek word for "hot") started out as a trademark in 1904, but it became so widely used that it's now a generic term for any vacuum container. One manufacturer still uses it as a brand name, however.

You didn't ask, but . . .

How does Styrofoam work as an insulator?

Unlike *thermos*, which has become a generic word, *Styrofoam* is still struggling to retain its identity as a trademark, but nobody seems to be paying attention; people call all polystyrene foam products "styrofoam" anyway.

The material is a good insulator because the plastic foam contains billions of trapped gas bubbles. Gases hinder heat conduction because their molecules are so far apart that they're very difficult for other molecules to collide with, either to give or to take away energy. The polystyrene plastic in between the bubbles is a good insulator also, because its molecules are so big that they can't move around much.

The thin Styro—I mean, polystyrene foam—boxes that restaurants pack your doggie-bag food in are supposed to keep the food hot on your way home. Instead of staying really hot, though, the food is probably being kept at just the right temperature for bacteria to flourish. Then, when you get home, you put the whole box in the refrigerator for the next day's lunch, but the foam insulation may keep it at maximum spoilage temperature for another hour or so. Better to transfer the food into an *un*insulated container before putting it in the refrigerator.

Freeze! I've Got You Uncovered

I took a can of soda pop from the refrigerator and the instant I opened it, it froze solid. What happened?

The soda wasn't frozen as long as it was still in the refrigerator because the refrigerator's temperature was warmer than its freezing point. But when you pulled the tab you did two things: You released the pressure inside the can and you lost some of the gas. For different reasons, each of these effects helped the liquid to freeze.

Every liquid has a certain temperature at which it will freeze: its freezing point. The freezing point of pure water is 32 degrees Fahrenheit or 0 degrees Celsius. Impure water—water that has any kind of stuff dissolved in it—has a colder freezing point than pure water does (see p. 92).

The more stuff is dissolved in the water, the colder its freezing point will be.

Soda pop certainly has a lot of stuff dissolved in it: sugars, flavors, and, especially, carbon dioxide gas. So it won't freeze until well below 32 degrees Fahrenheit. But as soon as you opened the can, the liquid lost some of its burden of dissolved carbon dioxide gas, which escaped from the liquid and went off into the air. Now containing less dissolved stuff, the liquid's freezing point becomes warmer than its own temperature from the refrigerator, and it dutifully freezes.

Opening the can and releasing the pressure had another effect as well. Ice occupies more volume than liquid water does (p. 202). So, if you compress ice it tends to revert to its smaller-volume liquid state; it melts. Under the high-pressure conditions in the closed can, the ice was repressed and remained liquid. But as soon as you released the pressure, the liquid water was free to expand into its higher-volume form: ice. Of course, this couldn't have happened unless the soda was already colder than its freezing point because it had already lost the gas.

As if that weren't enough, there was a third effect. When you opened the can, the compressed carbon dioxide gas was able to expand. Whenever a gas expands, it cools (see p. 135). This extra cooling also contributed to the freezing.

Either turn your refrigerator down—that is, turn the temperature up—or don't open the cans until they've warmed up a bit. You can wait.

Getting Hot in Bed

Why do water beds have to have heaters? A few days after filling one, won't the water be just as warm as anything else in the room, including any other type of bed?

The water in a water bed will indeed settle down to the same temperature as everything else in the room, including a conventional bed. But you would still *feel* colder on the water bed. It has to do with the fact that water conducts heat away from your body a lot more efficiently than other materials do, such as a conventional mattress.

Heat is nothing more than the motion of a substance's molecules (see p. 236). Various materials can transmit that motion, and hence conduct heat, with varying degrees of efficiency. The best way is by *conduction*—transmission directly from one molecule to the next to the next, and so on down the line. In order to do this, adjacent molecules must be close enough together to poke elbows.

In water, the molecules are just about touching, so the faster-moving ("hotter") molecules can easily transmit some of their motion to the adjacent "cooler" molecules. The heat—in this case, your body heat—thus travels efficiently into the water, and you'll feel cold unless some of that heat is restored to you by an electric heater.

Mattresses are a much poorer conductor of heat than water because they contain air. In air, the molecules are very far apart, with lots of empty space between them (see p. 151). They can therefore bump against each other only

rarely, so the transfer of heat motion can take place only rarely and heat transmission is very slow. On a regular mattress, your body is putting out heat faster than the mattress can lead it away, so you stay cozy.

Want to be *really* cold? Try sleeping on a metal slab. Metals are superb conductors of heat because their atoms are held very close together by a "cement" of electrons.

TRY IT Try thawing two boxes of frozen strawberries, one by leaving it out in the air at around 75 degrees Fahrenheit, and the other by immersing it in a bowl of cool tap water at around 65 degrees. Even though the water is cooler than the air, the strawberries will thaw faster in the water because water conducts heat to the box—that is, removes cold from the box—more efficiently.

BAR BET Frozen strawberries can be thawed faster at 65 degrees Fahrenheit than at 75 degrees Fahrenheit.

The Cigarette-Smoke Blues

I've heard that back in the Dark Ages when people smoked cigarettes, the smoke rising from the cigarette was blue. But after the doomed one inhaled the smoke and blew it out again, it was white. I know what probably happened to the lungs, but what happened to the smoke?

Tar and nicotine are not blue, so forget that idea. What happened was that the size of the smoke particles had changed.

The particles in cigarette smoke as it rises from a quietly burning cigarette are extremely tiny, smaller than the wavelengths of visible light. When a passing light wave encounters one of these tiny particles, the particle is too small to bounce the wave backward like a handball from a wall. Instead, the wave is merely deflected somewhat from its path and continues off at an angle: It is *scattered.* The shorter wavelengths of light—at the blue end of the visible light spectrum—are scattered more out of their original paths than the longer wavelengths, because they are closer in size to the smoke particles.

When we look at the smoke with the main source of light behind us or off to one side, many of the blue rays aren't going straight through and being "lost" to us; they're being scattered around the room—more so than the other colors. Thus, our eyes receive an excess of bounced-back blue light and the smoke appears bluish.

When a cigarette is puffed upon, the smoke particles are somewhat larger, because they don't get a chance to burn down completely. When inhaled, many of them get trapped in the lung, where they are not seen again until the biopsy.

Those particles that do complete the round trip to lungland come out coated with moisture, which further increases their size. The particles are now bigger than the wavelengths of all colors of light, and they therefore don't scatter any of it. Like any large object, they reflect all colors equally, right back to where they came from. The smoke therefore doesn't appear to have any particular color and it looks white.

You didn't ask, but . . .

No science book can be complete without answering the question, Why is the sky blue?

It's blue for the same reason that cigarette smoke is blue: the preferential scattering of blue light by tiny particles.

Pure air is colorless, of course, meaning that all visible wavelengths (colors) of light pass through it without being absorbed. But it contains molecules and, often, suspended dust motes that are smaller than the wavelengths of visible light and that therefore scatter it. As is the case with the cigarette smoke particles, the blue light is scattered more than the other colors, which tend to go straight through the air without much change of direction.

When you look at the sky, you're seeing all the colors in sunlight that are coming down toward you, mainly from some direction off to one side—wherever the sun happens to be. But in addition to that, you're getting some extra blue light that is being "scattered off the air" from many other directions. Thus, you're receiving an excess of blue light over what the sun is putting straight out, and the sky looks bluer than the sun's own daylight.

You didn't ask this either, but . . .

Why are sunrises and sunsets so colorful?

When the sun is low in the sky at sunrise or sunset, you're seeing it straight-on through a great distance of atmosphere (p. 145). While traversing all that atmosphere, a lot of the blue light that started out in your direction gets scattered into many other directions, so the light that reaches you straight-on is depleted in blue. Sunlight that is depleted in blue looks red, orange or yellow, depending on what size particles of dust happen to be in the air, and what other colors they are therefore scattering.

If that kills the romance, forget that I ever said anything.

TRY IT Make your own sunset. Add a few drops of milk to a clear glass of water and look through the glass at a light bulb. The bulb will look red, yellow or orange. The light coming to you from the bulb is depleted in blue because of scattering from the tiny casein particles and butterfat globules suspended in the milk. The exact color that you see depends on the size and concentration of these particles in the water.

The Fizz-ics of Champagne

Why does shaking a bottle of soda or beer make it explode when you open it? And does opening a bottle of champagne have to be so messy? After all, when it hoses out the candles, some of the mood is inevitably lost.

The trick, as you already suspected, is to chill the bottles well and avoid any agitation for at least several hours before opening. But knowing why always helps.

Beer, soda, and champagne all get their fizziness from carbon dioxide gas, which has been dissolved in the liquid during—or in the case of true champagne, after—the bottling process. The fizz consists of bubbles of carbon dioxide coming out of the liquid into the air. When that happens gently on our tongues, we get that nice tingly sensation. But when it happens too fast, we get the mop.

The amount of carbon dioxide that can remain peacefully dissolved in a liquid depends directly on how much carbon dioxide there is in the space above the liquid's surface, because the more carbon dioxide molecules there are bouncing around in that space, the more of them will hit the surface and dissolve.

In the sealed bottle, that space is filled with carbon dioxide and air; furthermore, these gases are packed in very tightly, at a pressure that can be as high as sixty pounds per square inch or 4.2 kilograms per square centimeter. (The air pressure in your auto tires is only about half that much.) So there is lots of carbon dioxide dissolved in the liquid when it comes from the bottler.

When the bottle is opened, no matter how gently, the pressurized carbon dioxide escapes and only normal air at normal pressure exists above the surface. In normal air, only about one out of every three thousand molecules is carbon dioxide. So practically all of the dissolved carbon dioxide has to come out of the liquid in one way or another. The only question is, how fast? And the answer is, it's normally quite a slow process.

After an initial burst of escapees from the air space, the dissolved carbon dioxide molecules don't leave the liquid all at once. If they did, your beverage would go flat as a shadow in an instant, making one hell of an explosion no matter how gently you had handled the bottle.

Neither can the gas molecules leave one at a time from deep within the liquid. They have to find some rallying points, some unique meeting places at which they can congregate and form groups—bubbles—that are big enough to muscle their way up and out of the liquid. Scientists refer to these congregation sites as *nuclei* (plural of *nucleus*).

Almost any break in the homogeneity of the liquid—even a microscopic speck of dust—will serve as a nucleus for the formation of bubbles. So will tiny scratches on the surface of the glass, because they can trap microscopic air bubbles when the beverage is poured, and these air bubbles will invite more gas molecules to join them. Carbon dioxide molecules congregate at all of these nuclei and grow into bubbles, which rise as soon as they are big and buoyant enough to push their way upward through the liquid.

What does all this have to do with shaking the bottle? Well, when you shake the bottle, you are trapping some of the gas that was above the liquid, making tiny bubbles out of it. And these tiny bubbles are the best possible nuclei for the further growth of bubbles. The carbon dioxide molecules in the liquid latch onto these new bubbles, which grow into bigger and bigger bubbles. Before you know it you have a foaming mess, propelled out of the neck of the bottle by expanding gas pressure like a pellet out of an air rifle.

You will run into the same problem, but probably not as bad, if you open beer, soda, or champagne that isn't chilled enough. Carbon dioxide is less soluble in warmer liquids (see p. 16) so more gas will rush out than if the liquid were cold. If you have also shaken the bottle a lot—well, that's just too awful to contemplate.

You didn't ask, but . . .

Why do dainty little bubbles form in a glass of champagne and rise in genteel little streams, while the bubbles in a glass of beer seem to blurp up from all over the place?

There are several reasons, none of them sociological.

- The champagne is likely to have been poured into a flute, a tall, narrow glass that doesn't have a lot of bottom surface for bubble formation. Moreover, such narrow glasses are not as likely to have been scratched on their inside surfaces because (a) scrubbing instruments can't get into them as easily and (b) they have probably been used less often than beer mugs. Fewer scratches means fewer nuclei, which means fewer and smaller bubbles. You will see them rising from only a select few nucleation sites.

TRY IT Scratch the inside of a glass of beer or champagne with the tip of a knife, and you will see new bubbles arising from that brand-new nucleation site.

• Champagne is clearer than beer. True champagne (it says *méthode champenoise* on the label), as opposed to cheap sparkling wine, has been carefully clarified by cooling, settling, and *dégorgement*, or disgorging. In that process, the corked bottles are tilted neck downward and rotated periodically over a long period of time; the neck is then frozen and the plug of frozen sediment is shot out with the cork. Less suspended matter in the liquid means, again, fewer nuclei for bubbles to grow on.

• The carbon dioxide in true champagne is made right there in the corked bottle by added yeast and sugar during an aging process that goes on for months and sometimes years. During that long time the yeast cells not only die, as they do in beer and other wines, but their proteins decom-

pose into fragments called peptides. Every peptide molecule has one end that is a base, which can grab onto a carbon dioxide molecule, which is an acid, thereby trapping it in the solution.

So champagne can not only hold more carbon dioxide than the other beverages, but it gives it up more reluctantly after the bottle has been opened. Hence, the tiny streams of aristocratic bubbles, rising in orderly, one-by-one fashion from whatever nuclei happen to be available.

If you stopper and refrigerate the bottle, a good champagne will still be fizzy the morning after. And even the morning after that, if you are really serious about celebrating.

Your Spoon's Immune

At a dinner party at a friend's house, I stirred my coffee and the spoon got very hot, seemingly even hotter than the coffee. That never happens at home. What's going on?

Congratulations. Your friends think highly enough of you to put out their company tableware, which is made of sterling silver. Your home "silverware" is either stainless steel or (sorry about that) only silver-plated base metal.

Sterling silver is almost pure silver: 92.5 percent, to be exact. And silver is the best conductor of heat among all the metals. Heat will always move from a place of higher temperature to a place of lower temperature if it can find any way to get there (see p. 22), and silver provides a superb heat highway. All the spoon did was to conduct the coffee's heat out of the cup and into the cooler room or—when you touched it—into your fingers.

During the process of being a conduit for all that heat, the spoon itself becomes hot—approximately the same temperature as the coffee, even though you might think it's hotter. (I don't recommend sticking your finger into the coffee to prove that.)

Stainless steel conducts heat less than one-fifth as fast as silver does. At home you probably never leave your everyday spoon in the coffee long enough for it to get very hot at the handle end. Even if you did, it wouldn't conduct its acquired heat into your fingers fast enough to be uncomfortable.

We All Scream for Ice Cream

My ice cream freezer uses an ice and salt slush to produce an extra-low temperature. How does salt make it so much colder than the usual temperature of ice water?

The normal temperature of an ice-and-water slush is 32 degrees Fahrenheit (0 degrees Celsius). But that's not cold enough to freeze ice cream. It has to be at 27 degrees Fahrenheit (–3 degrees Celsius) or below. Salt is what does the job. Lots of other chemicals would do the trick, but salt is cheap.

When ice and salt are mixed, some salt water is formed and the ice spontaneously dissolves in the salt water, making more salt water. That's what happens when you throw salt on an icy sidewalk or driveway; solid ice plus solid salt becomes liquid salt-water (see p. 96).

Inside a piece of ice, the water molecules are fixed in a definite, rigid geometric arrangement (see p. 202). This rigid arrangement breaks down under the attack of the salt, and the water molecules are then free to move around loosely in the form of a liquid.

But it takes energy to tear down the solid structure of ice molecules, just as it takes energy to tear down a building (see p. 122). For a piece of ice that is in contact with nothing but salt and water, that energy can come from nowhere but the heat content of the salt water. So as the ice breaks down and dissolves, it borrows heat from the water, lowering its temperature. The slush gets repaid by taking heat out of the ice-cream mixture, which is, of course, just what you want it to do.

TRY IT Put equal amounts of cracked ice in two identical glasses. Pour just enough water into each one to make the ice begin to float. Then dump a lot of salt into one of the glasses and poke it down into the ice a bit. After several minutes, check the temperatures with a kitchen meat thermometer. (Lord knows why, but many of them do go down below freezing.) You'll find that the salted ice gets much colder than the plain ice. You may even be able to scrape some frost off the outside of the salted glass with your fingernail.

Some Like It Hotter

It's infuriating! Whenever I'm washing my hands or, worse yet, when I'm taking a shower, I carefully mix the hot and cold water to get just the right temperature. Invariably, just as I'm getting comfortable, the water gets colder and I have to mix it all over again. Is there a scientific, rather than a paranoid-schizophrenic, answer to this?

Yes, and a very simple one. Heat makes things expand. In a compression faucet (the most common kind), the water flows through a narrow gap between a neoprene rubber washer and a metal "seat." In the hot-water faucet, the initial flow of hot water makes the washer expand, which closes down the gap between washer and seat, restricting the flow of water. With less hot water flowing than you originally selected, the mixture is now colder.

There are several things you can do:

1. Replace the neoprene washer in the hot faucet with a "sandwich" type: fiber composite on the outside and rubber

on the inside. The fiber doesn't expand and contract as much as rubber does.

2. Don't be so stingy with the hot water. If you open the faucet wider, the slight constriction due to expansion won't even be noticeable. Of course, to get the temperature you want, you'll have to open the cold water faucet wider also.

3. Preheat the hot water faucet parts by running the water for several seconds after it flows hot. Then when you adjust the temperature, the diabolic expansion will already have taken place.

4. Take cold showers.

Or, for a change of pace, ask your live-in to flush the toilet while you're in the shower. You'll get all the heat you want. Fast.

What Goes Up Won't Come Down

The mercury in my fever thermometer seems to have no trouble going up, often higher than I'd prefer, and staying there. But to get it back down again I have to shake the thermometer hard. If the mercury went up so easily, why won't it come down?

If you look carefully, you'll see a narrow constriction in the capillary tube through which the mercury travels up and down. On its way up, the mercury has plenty of force to overcome the resistance and push through the constriction; the pressure of an expanding liquid can be quite high. (When water freezes and expands, the pressure can crack iron pipes and concrete walls; see p. 202.)

When you remove the thermometer from your mouth and the temperature of the bulb goes down, the mercury thread doesn't flow back down; it remains at its highest level. The mercury is certainly contracting in the bulb, but it can't pull the entire thread down along with it because the thread simply isn't strong enough; the attractive forces between mercury atoms are too weak to withstand the pull. (If they were much stronger, mercury would be a solid instead of a liquid.)

So instead of being pulled down into the bulb, the mercury thread breaks at the constriction like a cotton thread breaking at its thinnest point. The lower pool of mercury continues to contract down into the bulb, leaving a space (actually a vacuum) between it and the mercury that has been stranded above the constriction, like a line of freight cars that have become uncoupled from the rest of the train.

When you shake the thermometer down, you're swinging it in a circular arc. Centrifugal force slings the mercury outward toward the edge of the circle, which is down into the bulb, thereby overcoming the friction at the constriction.

Why Do Batteries Die?

Almost everything these days is powered by batteries. What's inside them? It must be electricity in some form, but how does it stay in there until we want some gadget to keep going . . . and going . . . and going?

Batteries do not contain electricity as such; they contain the *potential* for electricity in the form of chemicals. These chemicals are isolated from one another inside the battery and are thus prevented from reacting until we hook up the battery to a device and turn on the switch. Then they react and produce electricity.

Getting energy from chemicals is nothing new. We get heat energy from wood, coal, and oil—chemicals all—by burning them; that is, by allowing them to react with the oxygen in the air. These combustion processes belong to a whole class of chemical reactions that can be made to give off electrical energy, instead of heat energy.

Chemists call them *redox* (REE-dox) reactions. They're

quite common. Every time you use laundry bleach, for example, there's a redox reaction going on in your washing machine (see p. 42). You don't see the electricity because it is internal to the chemicals that are reacting. It is absorbed by certain atoms as fast as it is being produced by others. A battery is just a clever device that controls the chemical reactions in such a way that we can tap off that electrical energy whenever we need electricity. But first, let's see exactly what electricity is.

An electric current is a flow of electrons from one place to another. But where do the electrons come from? Electrons are everywhere; they are the outer parts of all atoms. So if we want an electron to move from one place to another, it has to leave atom A and skip to atom B, like a flea jumping from dog to dog. In order for this to happen, however, atom A must be willing to give up one of its electrons and atom B must be willing to take it on.

Different kinds of atoms have different degrees of affinity for their electrons. Some atoms actually try to get rid of an electron or two whenever they can, while others hold onto their electrons tightly and will even try to capture more. When an atom of the first kind (atom A) meets an atom of the second kind (atom B), they can make a mutually beneficial deal by transferring an electron or two from A to B. And that, in a nutshell, is what happens in a redox reaction.

This electron-passing game from one atom to another constitutes a flow of electricity on a microscopic, one-atom-at-a-time scale. The problem from our human-sized point of view is that if we try to get a usable amount of electricity by mixing a zillion atoms of type A with a zillion atoms of type B, the electron-passing takes place from atom to atom in all directions, in one big chaotic scramble, wherever an A can find a B. That is of absolutely no practical use to us.

We need those electrons to be passed from a large group of A atoms in one location to a separate group of B atoms in another location, through a one-way street, or circuit, that we provide. Then, in their eagerness to get from the A's to the B's, those electrons will have to push their way through our circuit, doing work for us along the way—anything from lighting a flashlight bulb to making a little pink bunny wander vacuously around while beating on a drum.

To make a battery, then, we'll make a compact little package containing lots of A atoms and lots of B atoms. But we'll keep them separated from one another, usually with a barrier of wet paper. They won't be able to do their electron passing until such time as we complete the circuit, when we hook up the battery and close a switch that allows the electrons to flow from the A atoms through our interposed gadgetry to the B atoms.

Different types of batteries are made from different kinds of A and B atoms. The most common ones are manganese, zinc, lead, lithium, mercury, nickel, and cadmium. In the familiar AAA (no relation to what we've called "A atoms"), AA, C, and D batteries (there once was a B battery, but it isn't used anymore), zinc and manganese atoms are the A's and B's. The zinc atoms are the electron passers and the manganese atoms are the electron receivers.

The battery's voltage, in this case 1.5 volts, is a measure of the *force* with which zinc atoms pass their electrons to manganese atoms. Different combinations of passer and receiver atoms will make batteries with different voltages, because they have different degrees of eagerness for passing and receiving electrons.

When all the passer atoms have passed their quota of electrons to the receivers, the battery is dead, and, alas, the bunny stops here.

Nicad (nickel-cadmium) batteries, as well as your automobile's lead-acid battery, are rechargeable, however: We can reverse the electron-passing process by pumping electrons back from the receivers to the passers, and then the passing game can begin all over again. Unfortunately, though, every time the battery is recharged, some mechanical damage is done to its innards, and even a rechargeable battery won't last forever.

You didn't ask, but . . .

Once a battery sends the electrons out into an electrical device, the electrons flow through the device and back to the battery again, don't they?

Not exactly. Inside the battery, electrons are indeed passed from one atom to another like jumping fleas. But that's not how electricity flows through a wire or through a complicated electric circuit. The electrons don't just enter one end of a wire, hop from one atom to the next, and come out the other end.

Let's say that the battery's voltage is pushing electrons through a wire from left to right. What really happens is that each electron repels its right-hand neighbor, because they are both negatively charged and similar charges repel each other. This nudges the neighbor toward the next right-hand neighbor, which nudges *its* neighbor, and so on.

By the time the wave of nudging gets to the other end of the wire—which is a lot faster than an electron can get there by broken-field running through the jungle of atoms—the effect is exactly the same as if those end-of-the-wire electrons were the original beginning-of-the-wire electrons. Who can tell one electron from another, anyway? Not even another electron.

Out, Damned Spot!

How does laundry bleach tell white from colors? Apparently, it can take any stain that humans don't like, no matter what its chemical composition might be, and turn it into white. How does bleach know what we want it to do?

Bleach knows nothing about white. What it does know about is color, because color is a lot more fundamental, chemically and physically speaking, than our mere human wash day preferences. Bleach attacks colored chemical compounds, most of which do indeed have something in common, and leaves behind a lack of color that we like to think of as "white."

Before I am attacked for calling white the absence of color when you learned in school that white is the presence of *all* colors, let me explain.

Light from the sun does indeed contain all colors of the rainbow—all the colors that humans can see, and then some. When all of these colors of light are combined, as they are in daylight, our particular brand of vision perceives the light as being no particular color at all. We call it white light.

But that's the light itself. What do we see when we look at an *object* that's being illuminated by that light? If the object reflects back to our eyes, equally, all the colors that fell upon it in the form of daylight, then the reflected light still has no apparent color to us—it's still white. We say that the *object itself* is white because we can only judge it by the light that it sends to our eyes.

If, however, the object has a particular appetite for, let's say, blue light, and it absorbs or holds back some blue parts of the daylight before reflecting the rest back to us, then the light we see will be deficient in blue. Our eyes perceive blue-deficient light as yellow, and we therefore call the object itself yellow.

If the "object" in question happens to be a stain on our otherwise-white (colorless) T-shirt, then the stain appears yellow to us, and we'll turn it over to our good old dependable laundry bleach for obliteration. We'll do the same when a stain happens to absorb some other specific color of light, thereby appearing to us as some other nonwhite color.

What is it, then, that the bleach is actually acting upon when it removes color? It is acting upon those molecules that have a preference for absorbing some specific color or colors of light, *any* specific color. The question then becomes, how does bleach attack only light-absorbing molecules?

When a substance absorbs light energy, it's the electrons

in the molecules that do the absorbing. By absorbing the energy, the electrons promote themselves to a higher level of energy-status within their molecules. The molecules of many substances are colored because they contain electrons that are of particularly low-energy status to begin with, and that are therefore eager absorbers of light energy. What bleach molecules do is gobble up these low-energy electrons so that they are no longer available to absorb light; thus, the molecules lose their coloring ability. (Techspeak: Electron gobblers are called *oxidizing agents*; the bleach *oxidizes* the colored substance.)

The electron gobbler that is commonly used in the laundry is sodium hypochlorite. Liquid bleaches are nothing but a 5.25 percent solution of that chemical in water. Powdered bleaches are usually sodium perborate, a gentler electron gobbler that doesn't attack most dyes. (Dyes, of course, are nothing but deliberate, tenacious, light-absorbing stains.)

Another popular electron gobbler, hydrogen peroxide, is used to bleach melanin, the dark coloring matter in human hair and skin. It is widely used in the manufacture of blondes.

2

In The Kitchen

There is no place in our daily lives where so many marvelously mysterious things are going on as in the kitchen. That's where we mix, heat, cool, freeze, thaw, and occasionally burn an incredible assortment of animal, vegetable, and mineral materials, using equipment that would have rattled the retorts and curdled the caldrons of the most devoted alchemist.

It's no accident that Shakespeare chose "fire burn and caldron bubble" as the most fundamentally mystical manipulations of his witches in *Macbeth.* Beneath the surface of these familiar operations, some extraordinary transformations are taking place—transformations of the sort that the alchemists could only have fantasized about, but that we can now explain in the simplest of terms—now that we know about the existence of molecules.

Do you think you already know what's going on during the boiling of a pot—or caldron—of water? Think again. We're going to start by taking a close look into that very pot to see what makes it tick . . . or bubble.

What's the Point of Boiling?

When I put a pot of water on the stove, I turn up the heat as high as it will go because I'm always in a hurry. But when it starts to boil, I have to turn the heat down to

avoid splashing. I still want the water to get as hot as possible, though, so it'll cook my stuff fast. Is there any way to make the water hotter without my having to mop up afterward?

Sorry, but once the water has begun to boil, it's as hot as it will ever get, even if you use a flamethrower. No matter how furiously you might get the water to boil, it won't get any hotter than the boiling point of water: 212 degrees Fahrenheit (100 degrees Celsius), plus or minus. (See p. 210 for a bit of quibbling about that "plus or minus.")

Let's take a close look at what's happening inside the water as you heat it to boiling. When you first start heating the water, its temperature rises. That is, the water molecules take on the heat energy and show it by moving faster and faster. Eventually, some of the molecules will have so much energy that they can actually break away from their buddies, to whom they had been attached by rather strong attractions. The energetic molecules may even elbow their buddies aside to make spaces in the liquid—bubbles, which then rise and erupt at the surface as blurps of water vapor—a gas. We can't see this gaseous water until it gets away from the surface, cools a bit, and condenses into a cloud of tiny liquid droplets that we call steam.

We refer to this whole complex process as "boiling." The bottom line is that the water is absorbing the heat that you're putting in and using it to change from a liquid to a gas.

Turning liquid water into gaseous water uses up heat energy, because it takes energy to break the molecules away from each other. If the molecules didn't stick together, water wouldn't be a liquid; it would always be a gas: loose molecules, flying around independently. Every liquid has its own degree of molecule-to-molecule stickiness, and therefore its own breakaway energy, and therefore its own boiling temperature. It requires a temperature of 212 degrees Fahrenheit (100 degrees Celsius) to break water molecules away from each other.

Now let's turn up the heat. The more heat energy per second we pump in from the fire, the more water molecules per second will be acquiring enough energy to break away and shoot off as gas. The water will boil more vigorously and it will boil away faster.

But all that extra heat *doesn't* make the temperature of the water go up, because any extra energy that a molecule may acquire, beyond what it needs to break away, simply goes flying off along with the molecule. As soon as any molecule gets more than the necessary breakaway energy, it will depart at a higher-than-usual speed. The extra energy, and the higher temperature that goes along with it (see p. 236), winds up in the steam, not in the liquid that remains behind in the pot. Its temperature will remain the same—at the "boiling point" of water—until the water has all boiled away.

Moral: You can't cook your spaghetti any faster by turning up the heat. Save your energy.

TRY IT You can check this for yourself by holding a candy or meat thermometer (with tongs) in the steam just above a pot of boiling water and in the water itself, both when it's boiling gently and when it's boiling vigorously. The water temperature will stay the same, but the steam will be a little hotter when the boiling is more vigorous.

BAR BET No matter how hot a fire you put under a pot of boiling water, the water won't get any hotter.

A Water-Gate Cover-up

I've noticed that a covered pot of water boils sooner than an uncovered one. I presume that the pot lid is keeping in some heat that would otherwise be lost, but what kind of heat? There is no hot steam to lose until the water is actually boiling, is there?

Steam, no; but water vapor, yes. Long before you ever see any steam, which is made up of tiny droplets of liquid water, there is a lot of invisible, hot vapor being produced—unattached water molecules, flying around as a gas. (A vapor and a gas are the same thing; people tend to call a gas a vapor if they know that it was recently in the form of a liquid.)

There is always some water vapor in the air above water, wherever it is. (You've heard of humidity.) That's because there are always some water molecules at the surface that happen to be moving vigorously enough to break away from their fellows and fly off. The higher the water's temperature, the more water vapor is produced, because more and more of the molecules will be moving vigorously enough—will be hot enough—to escape. So as the stove heats the water, the number of hot water-vapor molecules in the air above the water increases.

Because the vapor molecules are progressively higher- and higher-energy ones as the temperature of the water goes up, it becomes more and more important not to lose them. The pot lid closes the gate, so to speak, keeping most of the vapor molecules from escaping and sending them, with their heat energy still intact, right back into the pot. Hence, the water will reach the boiling point faster.

Unless, of course, you watch the pot.

Too Many Ions in the Fire

I've read that when you add salt to boiling water, it makes it hotter. Sounds impossible to me, but if that really does happen, where does the extra heat come from?

It's strange, but true. The boiling water will indeed begin to boil at a higher temperature as soon as the salt dissolves.

For every ounce of salt that you add to a quart of water (or for every 29 grams of salt per liter of water), the boiling temperature will increase by about 0.9 degree Fahrenheit (0.5

degree Celsius). That's no great shakes, but it's an increase nevertheless. Because the temperature-raising effect is so small, adding salt to your spaghetti water isn't going to cook it noticeably faster. You're adding the salt mostly for flavor, but some people claim that it gives pasta a firmer texture.

TRY IT *With a meat or candy thermometer, test the boiling temperature of a quart of water in a saucepan. Now add six ounces (about half a cup) of salt to the water and stir to dissolve. After the salt is dissolved and the water comes back to a boil, the boiling temperature will be about five degrees higher than it was.*

The "extra heat" that causes the higher temperature obviously can't be coming from the room-temperature salt that you added. But the burner on the stove is putting out lots more heat than the water really needs to boil. (You can feel it all around, can't you?) So there's plenty of heat available for the water to use if it chooses to increase its boiling temperature. The real question is, why does it choose to do so?

A liquid will boil when its molecules get enough energy to break away from each other and go flying off into the air. When salt (sodium chloride) dissolves in water, it splits up into electrically charged sodium and chlorine particles. (Techspeak: sodium and chloride *ions*.) These charged particles do two things:

First of all, they crowd the water molecules, hindering their ability to muscle their way out of the drink and fly off. It's as if the water molecules were people trying to get out of a bus by elbowing through a suddenly developed crowd. What they need is an extra shove, an extra amount of energy to help them escape. They therefore require a higher temperature for the boiling process.

The second thing that the charged sodium and chloride particles do is that they gather around themselves clusters of water molecules, which they wear like bulky little wet suits wherever they go. Charged particles can attract water molecules because the water molecules themselves are charged: slightly positively at one end and slightly negatively at the other (p. 66). (Techspeak: The water molecules are *polar*.) Their positive ends are attracted by the negative chloride particles and their negative ends are attracted by the positive sodium particles.

As a result of all this clustering, the sodium and chloride particles essentially remove from circulation a large number of water molecules. In order for these clustered water molecules to boil off, they've got to break away from the sodiums and chlorides—out of the wet suit—and that's more difficult than simply breaking away from their fellow water molecules if the salt weren't there. Hence, a higher boiling temperature is needed.

There's nothing unique about salt, however. Dissolving *anything* in water—sugar, wine, chicken juices, you name it—will produce the same hindrance effect, if not the same clustering effect. So don't ever say that your chicken soup is boiling at 212 degrees Fahrenheit (100 degrees Celsius) just because that's the number you learned in school for pure water. It's somewhat higher because of all the dissolved stuff in the soup.

Anyway, pure water boils at 212 or 100 degrees only when the weather is just right (see p. 210). And if there happens to be lots of sugar in the water, even stranger things can happen (see p. 53).

A Watched Pot Only Simmers

Why do recipes for stews and ragouts always warn me to simmer them, but not to let them boil? What's the difference, anyway? Isn't a simmer just a slow boil?

Not exactly. The difference between simmering and boiling is more basic than just the vigor of the bubbling. A simmer aims to produce a slightly lower temperature than true boiling, because even a few degrees difference in cooking temperature can make a big difference in how things cook. In moist cooking—cooking with lots of water present, as opposed to roasting—there is only a narrow range of possible temperatures, so getting exactly the right temperature can be tricky.

Cooking is, after all, a series of complex chemical reactions, and temperature influences all chemical reactions in two ways: It determines which specific reactions are going to take place, and it determines how fast they'll go. Everybody knows the general effect of temperature on cooking speed: the higher the temperature, the faster the cooking. But it's also true that different things happen to food when it is cooked at even slightly different temperatures, because different chemical reactions may be going on.

The question of temperature is particularly important in the water-cooking of meats. Meats undergo different tenderizing, toughening, and drying-out reactions (even when they're immersed in sauces) at different temperatures. The temperature of a full boil, for example, encourages the toughening process, but the slightly lower temperature of a simmer promotes tenderizing. Long experience has taught us which methods work best with which dishes, so it's wise not to monkey with recommended cooking methods.

Outright boiling—lots of big bubbles, such as in cooking pasta—is an infallible indicator of one specific temperature: the boiling point of water. This sets a definite upper limit to the temperature at which we can water-cook food, because water can never get above its boiling temperature, no matter how vigorously we boil it (see p. 45). All those bubbles bursting at the surface are telling us quite clearly that our food is cooking at just about 212 degrees Fahrenheit (100 degrees Celsius), depending slightly on various conditions (see pp. 48, 208, and 210).

But many desirable cooking reactions take place at lower

temperatures. How low? It depends on the food. The only important lower limit for cooking is the temperature that's needed to kill most germs: about 180 degrees Fahrenheit (82 degrees Celsius). The problem is, how do we achieve one of these lower cooking temperatures reliably when we need it? There is no obvious sign like bubbling to watch for, and we can't be expected to keep sticking thermometers into our pots all the time.

When cookbook writers want to tell you that something should be cooked at a certain temperature below actual boiling, they invoke words such as simmer, gentle simmer, slow boil, poach, and coddle. Then they wave their arms around, trying to describe what those words are supposed to mean. And they fail miserably. Look up "simmer" in professional books on cooking technique, and you'll be told that it means everything from 135 degrees Fahrenheit (Good luck! Dangerous salmonella bacteria aren't killed until 140 or 150 degrees), all the way up to 210 degrees; that is, anywhere from 57 to 99 degrees Celsius.

But trying to specify a standard "simmering temperature" is silly anyway, because the temperature inside a pot on a stove will vary quite a bit from one spot to another and from one moment to the next. A few of the factors that affect the food's temperature are the size, shape, and thickness of the pot; what it's made of; whether or not it is covered and, if so, how tightly; the steadiness of the heat source; the contact between the pot and the burner; the amounts of food and liquid in the pot; and the characteristics of the food itself.

There is only one way to achieve a proper simmer: Forget about temperature and concentrate on what the stew is *doing*. Carefully adjust the pot, the cover, and the burner so that bubbles are reaching the surface *only occasionally*. That means that the *average* temperature in the pot is somewhat below boiling, which is just where you want it. The occasional hot spots here and there throw occasional bubbles to the surface, just to let you know that it's not too cool.

Remember that real boiling is when almost *all* of the bubbles reach the surface. If the temperature is somewhat lower than the normal boiling point, bubbles may form at the bottom, but they are reabsorbed before ascending all the way to the surface. That's what's going on during a proper simmer.

What about poaching and coddling? Poaching is just another word for simmering, usually applied to fish or eggs. Coddling is placing the food, usually an egg, in water at boiling temperature and then turning off the heat. The temperature decreases steadily as the water cools, so that the average temperature turns out to be the kindest and gentlest of all. The result is a thoroughly pampered, humored, and overindulged egg.

Candy Is Dandy, but Heater Is Sweeter

How come sugar syrup gets hotter and hotter the longer you cook it, but plain water doesn't? Or does it?

You've been making candy, haven't you?

Candy recipes tell you to boil sugar syrup until it reaches various temperatures on a candy thermometer: the soft ball stage at about 237 degrees Fahrenheit (114 degrees Celsius), the hard crack stage at 305 degrees Fahrenheit (152 degrees Celsius), and so on. (Different cookbooks will give you slightly different temperatures for the various stages.) The longer you boil it, the thicker the syrup gets and the higher the temperature becomes. But you can boil pure water as long and as hard as you like, and it will never get any hotter (see p. 90).

Obviously there's something going on in that bubbling sugar-water syrup that's very different from what goes on in plain boiling water.

Whenever something—almost anything—is dissolved in water, the boiling temperature goes up. And sugar is no exception. So any sugar-and-water solution will boil at a higher temperature than plain water will. The more concentrated the solution or, put another way, the more dissolved material the water contains, the higher its boiling temperature will be (see p. 48).

For example, a solution of two cups of sugar in one cup of water (yes, it's possible; see p. 76) won't start boiling until 217 instead of 212 degrees Fahrenheit (103 instead of 100 degrees Celsius). But then as you continue to heat it, many of the water molecules will boil off as vapor, and the sugar solution will become more and more concentrated. It will develop a higher and higher proportion of sugar to water. The more concentrated it becomes, the higher its boiling temperature becomes, so the longer you boil it, the hotter it gets. Because of this, candy recipes can use temperature as an indication of how concentrated the sugar syrup is and therefore how hard or sticky it'll be when cooled.

If you boil sugar syrup long enough, all the water will eventually be gone and you'll be left with nothing but melted sugar in the pot, at about 365 degrees Fahrenheit (185 degrees Celsius). At about the same time, it will begin to *caramelize*, a polite word for the actual destruction of sugar molecules into a complex assortment of other chemicals that have intriguing flavors in spite of their generally frightful chemical compositions. The yellow-to-brown color transition corresponds to the buildup of more and bigger particles of carbon, which is the ultimate decomposition product of sugar. Heat sugar just a little too long, though, and you'll be left with a charred, black mess of still-sweet, but thoroughly inedible, charcoal.

Eggs Over, Not So Easy

The longer I cook my egg, the harder it gets. The longer I cook my potato, the softer it gets. Why does heat have such different effects on foods?

The short answer is that cooking makes proteins harder and carbohydrates softer. We'll leave meats out of this, because the toughness or tenderness of a cut of meat depends in a very complex way on the muscle structure of the animal (see p. 56), the portion of the animal it came from, and on precisely how it's being cooked. During cooking, for example, meat can get tenderer at first and then tougher later on. But your egg and potato can be explained entirely by the differing effects of heat on protein and carbohydrate.

First, let's take a close look at that egg. Eggs are rather unusual in their composition, as befits their unique function in life. If we throw away the shell of a hen's egg and remove the water from what's inside, the dried remains are just about half protein and half fat, with virtually no carbohydrate. The dried yolks are 70 percent fat while the dried whites are 85 percent protein. You know that heat doesn't affect the consistency of fat very much, so we'll concentrate on the protein in the egg white. And you also know that we're not going to get out of this without looking at what the molecules are doing, right?

The albumins in albumen—no kidding; egg whites are called *albumen*, but they contain proteins called *albumins*, with an *i*—are made up of long, stringy molecules coiled up into globs like very loose balls of yarn. When heated, these balls partially unravel and then stick to each other here and there, making an unholy tangle like a can of spot-welded worms. (Techspeak: The molecules become *cross-linked*.)

Now, when the molecules of a substance change from a bunch of loose balls to an unholy, welded tangle, the stuff is obviously going to lose its fluidity. It will also turn opaque, because even light can't get through it.

Liquid egg albumen, when heated above approximately 150 degrees Fahrenheit (65 degrees Celsius), coagulates into a firm, white, opaque gel. The hotter and longer you heat them, the more the molecules will unravel and spot-weld to each other. So the longer you cook an egg, the firmer its white will become, ranging from the glop of soft-boiled to the rubber of hard-boiled to the leather of an "over, well" hash-house special. The drying out that takes place at higher temperatures also contributes to the toughening.

The protein in the egg's yolk coagulates in much the same way, but not until it reaches a somewhat higher temperature. Also, the yolk's abundant fat acts as a lubricant between the globs of protein, so they can't weld together as much and the yolk doesn't get as tough as the white, even when hard-boiled.

Now, about that potato and other foods that contain a lot of carbohydrates. Starches and sugars cook easily. They even dissolve in hot water to speed the process. When you bake a potato, some starches dissolve in the steam.

But there's one very tough and very insoluble carbohydrate that is present in all of our fruits and vegetables: cellulose. The cell walls of plants are made of cellulose fibers held together by a cement of pectin and other water-soluble carbohydrates. This structure is what makes vegetables like cabbage, carrots, and celery—and potatoes—so firm and crisp. But put the heat on these tough guys, and they turn into wilted wimps. The pectin cement dissolves in the liquids released by the heat, and the rigid cellulose structure is severely weakened. The result is that cooked vegetables are softer than raw vegetables.

Fish: The Real White Meat

Why is fish flesh generally white, while most other meats are red? Fish have blood too, don't they? And why does fish cook so quickly compared with other meats?

Well, it's not because the fish has been marinating all its life. Fish flesh is inherently different from the flesh of most walking, slithering, and flying creatures for several reasons.

First of all, cruising through the water doesn't exactly

qualify as body-building exercise, at least when compared with galloping across the plain or flitting through the air. So fish muscles are not as developed as those of other animals. Elephants, for example, have to work so hard just to move against gravity that their highly-developed muscles are extremely tough and, as you undoubtedly know, must be simmered for a long time before they are tender.

But more important is the fact that fish have a fundamentally different kind of muscle tissue from most land animals. To dart away from their enemies, fish need quick, high-powered bursts of speed, as opposed to the long-haul endurance that most other animals must possess for running. So fish muscles are made predominantly of what are known as fast-contracting fibers. (Muscles are generally made up of bundles of fibers.) These are shorter and thinner than the big, slow-contracting fibers of most land-animal muscles, and are therefore easier to tear apart, such as by chewing, or to break down chemically, such as by the heat of cooking. Fish is even tender enough to eat raw, as in sashimi, but steak tartare has to be ground to render it vulnerable to our omnivorous molars.

Another big reason that fish are more tender than other animals is that they operate in an essentially weightless environment (see p. 186). They therefore have little need for connective tissue—the cartilage, tendons, ligaments, and such that other creatures need for supporting their various body parts and for fastening them to the skeletal tree. So fish are practically pure muscle, with none of those tough materials that have to be cooked into submission.

For these reasons, fish flesh is so tender that the main problem is to keep from cooking it too much. It should be cooked only until the protein becomes coagulated and opaque, pretty much the same as what happens to the protein in the white of an egg (see p. 54). They'll both get tough and dry if you cook them too long.

And why is fish flesh white? Fish don't have very much

blood, to be sure, and the small amount they do have is largely concentrated in the gills. But by the time any animal food gets to our tables, almost all of the blood is gone anyway. The answer has to do (again) with the different type of muscular activity in fish. Because their fast-contracting muscle tissue operates only in short bursts, it doesn't need to store up oxygen for endurance activities. Other animals' slow-contracting muscles must store up endurance oxygen, and they do it in the form of *myoglobin*, a red compound that turns brown when exposed to air or heat. It's the myoglobin, not the blood, that makes red meat red (see p. 127).

How to Stop Smoking

What's going on when I clarify butter? And why should I bother?

You're doing it to get rid of everything but the (yum, yum) pure saturated fat.

Some people think of butter as a block of fat surrounded by guilt. But guilt or no guilt, it isn't all fat. It's a three-part, solidified emulsion—a stable mixture of oily and watery components, with some solids mixed in. When you clarify butter, you're separating off the fat and throwing away everything else. Your objective is to be able to fry at a higher temperature than usual without any burning and smoking, because the watery stuff holds the temperature down and the solids do indeed tend to burn and smoke.

When heated in a frying pan, whole butter will start to smoke at around 250 degrees Fahrenheit (121 degrees Celsius), and the solid proteins in it will begin to char and turn brown. One way to minimize these happenings is to "protect" the butter in the pan with a little cooking oil that has a higher smoking temperature. Better yet, you can use clarified butter. The butter oil itself won't smoke until about 350 degrees Fahrenheit (177 degrees Celsius), and there will be no solids to scorch. The only problem is that a lot of the buttery flavor is in the casein and other protein solids that you throw away.

TRY IT To clarify butter, all you have to do is melt it very slowly over the lowest possible heat. (Remember, it scorches easily.) The oil, the water and the solids will separate into three layers: a froth of casein on the top, the oil in the middle, and a watery sediment of milk solids (whey) on the bottom. Just skim off the froth and decant the oil into another container. If it bothers you to throw away the flavorful casein froth, save it and use it to flavor vegetables.

Another reason to clarify butter is that bacteria can work away at casein and whey, but not at pure oil, so clarified butter will keep much longer than whole butter will. In India, they keep their *ghee*, which is thoroughly clarified butter, for long periods of time without refrigeration. Eventually, though, it'll turn rancid because air oxidizes the unsaturated fats in it. But rancidity is only a sour flavor, not bacterial contamination. Tibetans prefer their clarified yak butter to be rancid.

You didn't ask, but . . .

What is the "drawn butter" that they serve with lobster and other seafood in restaurants?

It's melted butter, plain and simple. It may be clarified in a way, by slow melting and pouring off the oil, or it may not. And it may even be margarine. Occasionally it may be thickened and seasoned.

Why "drawn"? In a sense, converting butter from a solid into a liquid can be thought of as "drawing it out." Besides, it looks impressive on the menu.

I'm Forever Baking Bubbles

There are two white, powdered chemicals in my pantry: baking soda and baking powder. I know they have different uses, but what are these chemicals, really, and what do they do?

The short answer is that baking *soda* is a single, pure chemical compound, while baking *powder* is a mixture of that same chemical with one or two others.

(If you must know, baking soda is sodium bicarbonate or bicarbonate of soda, while baking powder is sodium bicarbonate plus an acid or two: tartaric acid or potassium hydrogen tartrate or monocalcium tetrahydrogen phosphate and, if it is "double-acting," sodium aluminum sulfate.)

A lot of chemicals to put in our food? Well, if you could see the chemical formulas of all the proteins, carbohydrates, fats, vitamins, and minerals in the food, you might stop eating altogether. Chemicals are like cowboys: there are black hats and white hats, and you've simply got to choose your friends and enemies intelligently.

So far, even the most contentious natural-food advocates haven't found anything to complain about in the chemical compounds that reside in your little can of baking powder. In the chemical names above, you may recognize sodium, potassium, calcium, phosphorus, sulfur, and aluminum—all not only harmless, but all (except possibly aluminum) essential to life. And all the carbon atoms, oxygen atoms, and hydrogen atoms in those chemicals pretty much turn into harmless carbon dioxide and water when they hit the oven's heat.

The key to all of this is the term *carbonate* in the sodium bicarbonate. A carbonate turns into carbon dioxide gas when provoked either by heat or by an acid. And that's our reason for using carbonates in baking: They leaven baked goods with carbon dioxide gas. *Leaven* comes from the Latin *levere*, meaning to make light or to raise.

Baking soda makes baked goods rise by producing millions of tiny bubbles of carbon dioxide inside the dough or batter, thereby frothing it up. As the froth begins to firm up under the influence of the oven heat, the bubbles become trapped. The result is a light, spongy cake or biscuit instead of a hard lump of dried-out flour paste.

Baking soda, or pure sodium bicarbonate, produces carbon dioxide gas by reacting with any acid ingredients that happen to be present, such as vinegar, yogurt, or buttermilk. It doesn't give off carbon dioxide all by itself until heated to 518 degrees Fahrenheit (270 degrees Celsius). On the other hand, the reaction with acid begins as soon as the ingredients are mixed; you can see the bubbles forming in buttermilk pancake batter even before you put it on the griddle.

TRY IT Add a little vinegar to some bicarbonate of soda in a glass and watch it froth up with copious bubbles of carbon dioxide. Vinegar is a solution of acetic acid in water.

Baking powder is baking soda with a dry acid already mixed in, so no other acids are needed in the recipe. As soon as baking powder gets wet, the two chemicals dissolve and react with each other to produce carbon dioxide.

Various chemical compounds are used for the acid in baking powder: monocalcium tetrahydrogen phosphate (usually called calcium phosphate on the labels), tartaric acid, and potassium hydrogen tartrate (cream of tartar) are the most common. To prevent these ingredients from "going off" prematurely in the can, they are diluted with a lot of cornstarch, which keeps them separated from each other until they dissolve in the mixing bowl. Also, they have to be zealously protected from atmospheric moisture by keeping the powder in a tightly closed container.

TRY IT *Add some baking powder to a little water and watch it froth up with bubbles of carbon dioxide. If it doesn't, your baking powder is dead because it became moist and spent its reaction slowly while in the can. Toss it out and buy some fresh.*

In most cases, we don't want our baking powder to release all of its gas bubbles as soon as we mix the dough or batter, before it has hardened up enough to trap the bubbles in place. So we make our baking powder "double acting," which means that it releases only a portion of its gas upon mixing, and the rest only when the temperature gets high enough in the oven. Double-acting baking powders (and most of them are, these days) usually contain sodium aluminum sulfate, which might be thought of as a high-temperature acid.

Baking is a very complicated business; a lot more chemistry is going on than just the leavening. Over the years, different leavening agents have been found to work best in different recipes, from pancakes to biscuits to a million kinds of breads and cakes. The exact times and temperatures at which the bubbles are most advantageously released have been carefully worked out by trial and error. So no substitutions in the recipes, please. Ironically, nobody likes a pancake that's flat as a pancake.

Meltdown in the Kitchen

Why can I melt sugar, but not salt?

Who says you can't melt salt? Any solid will melt if the temperature is high enough. Lava is molten rock, isn't it? If you want to melt salt, all you have to do is turn your oven up to 1474 degrees Fahrenheit (801 degrees Celsius), which will make your kitchen glow a pretty red color. Ovens don't melt until around 2700 degrees Fahrenheit (1480 Celsius).

Of course, what you mean is that sugar melts much more easily than salt does—that is, at a much lower temperature. Sugar will melt at only 365 degrees Fahrenheit (185 degrees Celsius). The question, of course, is why? What's so different about these two common, white, granulated kitchen chemicals? They're both pure chemical compounds and may look similar, but they're members of two very different chemical kingdoms.

There are more than eleven million known chemical compounds, each with unique properties. In an effort to make sense out of this vast variety of substances without going completely mad (it has worked for most of them), chemists begin by dividing them into two broad categories: organic and inorganic.

Organic compounds are those that contain the element carbon. They are found mostly in living things or in ex-living things such as petroleum and coal.

Inorganic compounds are, well, everything that's not organic. Most foods, drugs, and living-thing chemicals—including sugars—are organic, while all the rocks and minerals—including salt—are inorganic.

If there is a single, sweeping generalization that can be made about the physical properties of organic and inorganic substances (and, of course, there are exceptions), it's this: Organic substances tend to be soft and inorganic substances tend to be hard. The reason for this is that the molecules that organic substances are made of are electrically neutral groupings of atoms, while the molecules that inorganic compounds are made of are usually *ions*: electrically charged groupings of atoms. The attraction of opposite charges for one another is a stronger force than the attraction of neutral molecules for one another; it's anywhere from two to twenty times stronger. Thus, inorganic substances are much harder to break apart—to separate the particles of—than organic substances are. You may have noticed that it's much harder to cut into a rock than into a tree.

Now what happens when something melts? It's really just like breaking the substance apart. The molecules start jiggling around so much from the heat that they eventually separate and begin to flow over and around each other, and the material turns into a flowing liquid. Obviously, loosely-bound organic molecules should be able to start flowing at a lower temperature because they don't require such vigorous agitation to unstick them from one another. So organic substances generally melt at lower temperatures than inorganic substances do.

Sugar (sucrose) is a very typical organic compound, consisting of neutral molecules. Salt (sodium chloride) is a very typical inorganic compound, made up of sodium ions and chloride ions. It should be no surprise, then, that sugar melts much more easily than salt does.

Like everything else, it's all in the molecules.

You didn't ask, but . . .

If every pure chemical substance has a specific temperature at which it melts from solid to liquid, does it also have a specific temperature at which it solidifies from liquid to solid?

Yes. As a matter of fact, the two temperatures are identical.

The solidification process you describe is what we also call *freezing*. When we say that water *freezes* at 32 degrees Fahrenheit (0 degrees Celsius), we could just as well say that that's the *melting* point of ice. The reason they are the same is that the slithering molecules of a liquid must be slowed down to a certain definite energy in order for them to fall into their

A calorie is not a Calorie

A calorie is an amount of energy. While energy can exist in a variety of interchangeable forms, the most familiar form to most people is heat. So a calorie is generally regarded as an amount of heat.

But precisely how much heat? Ask a chemist and you'll get one answer, but ask a dietitian or nutritionist and you'll get another. And they're not even close; one "calorie" is a thousand times bigger than the other. It's as if one person's kilometer were another person's meter; in order to interpret a highway sign, you'd have to know who wrote it.

There's no indication that the chemists and nutritionists will ever agree; they're both too set in their ways. So the world is stuck with two sizes of calories.

The chemist's calorie, which we might call a *gram calorie*, is the amount of heat it takes to raise the temperature of one gram of water (about 20 drops) by one degree Celsius. But that's a pretty small amount of energy, so the nutritionist uses the *food calorie*: the amount of heat it takes to raise the temperature of a *thousand* grams of water by one degree Celsius. Thus, one food calorie is equal to a thousand gram calories.

To avoid confusion in this book (good luck!), I'll use *calorie* with a lower-case *c* for the gram calorie and *Calorie* with a capital *C* for the food calorie. So when you see it written both ways, they're not typos. Anyway, you can pretty much tell which one is meant by its context.

permanent, rigid places in a solid crystal; on the other hand, they must be heated up to that same amount of energy in order to break free from those rigid positions and begin to flow as a liquid.

Thus, a certain definite amount of heat is involved in the melting-freezing transition of any substance between its solid and liquid forms. For pure water, that amount of heat happens to be eighty calories per gram. (See box.) If you want to melt a gram of ice, you've got to put eighty calories of heat into it; if you want to freeze a gram of liquid water, you've got to take eighty calories of heat out of it.

Just to be contrary, chemists do not call that amount of heat the "heat of melting" or the "heat of freezing." They call it the "heat of fusion." To make things even worse, whenever a substance happens to be a liquid at room temperature and we have to cool it to make it a solid, people call the transition temperature a freezing point, whereas if the substance is a solid at room temperature and we have to heat it to convert it to a liquid, people call that very same transition temperature a melting point. Go fight City Hall.

Rubbing a Molecule the Wrong Way

I would just love to know what my microwave oven is doing during all that humming and whirring. Is it really cooking the food from the inside out? Also, it says in several cookbooks that microwaves make food molecules rub up against each other, and that the friction makes them hot. But I'm not sure that cookbooks are the best place to learn science.

You have good reason to be suspicious. Both of these notions are misguided.

What is friction? When you rub the surfaces of two objects together, they'll resist slipping and sliding to some extent, and some of the muscle energy that you're applying to overcome that resistance shows up as heat. And what is heat? It's just molecular motion. What microwaves do is make certain molecules move, and once they're moving, they're hot, period. No rubbing is involved—whatever on earth "rubbing" is supposed to mean for a molecule. What do you rub it with? An atom on a stick?

What about cooking from the inside? Sure. Let's take three hundred pounds of ground beef, shape it into a giant meat loaf and tuck our oven into the middle of it, leaving the cord hanging out. Then we'll plug it in and push the ON button with a broom handle. That's the only way we might get a microwave oven to cook food from the inside.

The heat in an ordinary oven does indeed have to work its way in from the outside of the food, cooking as it goes; that's why the center of a roast is rarest. The heat has to penetrate by conduction—hot, fast-moving molecules banging into slower-moving molecules—and that can be quite a slow process; meat and potatoes are pretty poor conductors of heat.

Microwaves also have to enter from the outside, but they penetrate instantly. They're *electromagnetic radiation*, like the various wavelengths of light and radio waves (see p. 218). In fact, they're nothing but very-high-frequency radio waves, oscillating at a frequency of about two billion cycles per second (two thousand megahertz), which is about twenty times the frequency of the FM radio band.

Microwaves will pass right through most materials without being absorbed until they encounter molecules that happen to be good absorbers of energy at that particular frequency. Usually, this means that they'll penetrate an inch or two (several centimeters) into a chunk of typical food, heating and cooking as they go—the more absorbing-type molecules in their path, the more heating.

What are these microwave-absorbing molecules? Water is by far the principal one, and since all food contains water, all food will absorb microwaves to some extent. Fats are another fairly good microwave absorber, with carbohydrates and proteins coming in a poor third and fourth. So the portions of your meal that are moister and fattier will get the hottest. It then remains for conduction to distribute that heat to the rest of the dish. That's why the zapping instructions often tell you to let the food stand for a few minutes before uncovering it; that lets the steam and hot fat distribute their heat throughout the dish.

There's something special about water and fat that makes them absorb microwave energy. Their molecules are what chemists call *polar*: they're not electrically uniform. The electrons in the molecules spend more time at one end of the molecule than at the other, and that gives the molecules a slight negative charge at that end and a slight positive charge (a deficiency of negative charge) at the other.

This makes them behave like tiny electric magnets, with two poles. When the microwaves oscillate, reversing their electric field a couple of billion times a second, these polar molecules are forced to line up with the field and then about-face and line up in the opposite direction a couple of billion times a second. This makes for some exceedingly lively molecules—that is, some exceedingly hot molecules. And while they're flip-flopping, they'll jostle their neighbors, polar or not, making them hot also.

The molecules of air (nitrogen and oxygen) and the molecules of paper, glass, ceramics, and all those "microwave-safe" plastics are electrically uniform. They're not polar, so they don't flip-flop and don't absorb microwave energy.

Metals are quite another story, however. They reflect microwaves like a mirror (radar waves are microwaves; airplanes and speeding autos reflect them), and will keep them bouncing back and forth inside the oven until the energy builds up to an alarming—even a spark-producing—degree. Except in the form of little pieces of thin foil, metal is a strict no-no in a microwave oven.

Oh, the humming and whirring? That's just a metal fan

blade that scatters the microwaves uniformly throughout the box.

Sectarian Sodium Chloride

Why do some recipes specify kosher salt? How does it differ from gentile salt?

It is, of course, entirely unnecessary to point out that salt is inherently nondenominational. While kosher salt comes from the sea and is certified at the factory as meeting the strict criteria of Jewish dietary laws, the rabbi's blessing has no more effect on flavor than does the priest's consecration of a communion wafer.

Kosher salt is exactly the same, chemically, as any other salt. It is pure sodium chloride, and like all salt intended for human consumption, must be greater than 97.5 percent pure, according to U.S. law. The only pragmatic difference between it and secular salt is in the size and shape of the grains, the kosher variety being coarser and usually flakier. Its main use is in the koshering process, which involves covering meat or poultry with a blanket of salt to purify it.

It's also useful for certain nonritualistic purposes, however, because of its coarser grains, and that's the only reason it is sometimes specified over regular table salt. Food "experts" who claim that kosher salt has a different flavor from regular salt should be politely asked to go pound it.

TRY IT *Take a look at some ordinary granulated table salt under a magnifying lens. Unless you've had a good course in chemistry, you'll be astonished at how regularly shaped the grains are. They're actually tiny cubes! You'll notice that most of them are somewhat the worse for wear because their sharp edges have been worn off in jostling against their neighbors, and some have been chipped so badly* that they're practically spheres. But you can tell that originally, they all tried very hard to be perfect cubes.

The cubic shape arises from the geometric arrangement of the sodium and chlorine atoms that make up the salt particles. For reasons that chemistry teachers can take half a semester to explain (if they explain it at all), and that I won't burden you with here, it happens that when sodium and chlorine atoms combine to form sodium chloride, they arrange themselves into a perfectly square pattern. (It has to do with their electric charges and their relative sizes.)

When zillions of sodium and chlorine atoms come together to make a three-dimensional salt crystal that's big enough to see, the shape of that entire crystal is going to reflect the square geometric arrangement of its individual atoms. A cube is simply a square in three dimensions, no?

While the compact cubic grains of ordinary table salt are just right for shooting smoothly through the holes in a salt shaker, kosher salt must have more "cling" to it for coating

meat in the koshering process. Even though the individual atoms inside are still arranged in a cubic pattern, the grains have a more irregular outer shape. They come out that way in the form of a crust on the surface of sea water when it is evaporated slowly. In kosher law, that's considered to be a more natural process than taking salt from a mine, dissolving it in water, and then evaporating the brine with coal or gas heat—which is how most other table salt is prepared.

TRY IT *If you look at kosher salt under a magnifying glass, you won't see cubes. The crystals will be irregular and flaky.*

Chefs often prefer to cook with kosher salt because it is easier to pinch with the fingers and throw into a pot, and they can feel exactly how much they're using. As soon as it dissolves in the food, however, the former size and shape of its grains become totally immaterial.

The Emperor's Salt

According to what I've read in food magazines, sea salt is highly preferable to ordinary table salt because (a) it is loaded with nutritious minerals, (b) it is unrefined and therefore more natural, and (c) it has a sharper, fresher flavor. How should I evaluate these claims?

(a) Baloney. (b) Baloney. (c) Baloney.

The sea salt that's sold in supermarkets and health food stores is no richer in minerals, no less refined, and no different in flavor from ordinary table salt. But you'll pay anywhere from four to twenty times as much for it. And it may not even come from the sea because manufacturers don't have to specify their sources and according to industry insiders, fibbing does occur. (In the case of kosher salt, a rabbi is present to keep them honest.)

Sea salt has long been the darling of natural-food faddists, who seem to require no evidence whatsoever before adopting the latest fervent conviction. But in recent years, otherwise reputable cookbooks and food magazines have become peppered, so to speak, with paeans of praise to the salt of the sea. When professional food writers begin to climb aboard a fundamentally shaky bandwagon, it is time to stop the parade. What we're seeing is a classic case of the emperor's new clothes. Admit that you can't tell the difference between sea salt and ordinary salt, and you're confessing that your palate is not only insensitive, but politically incorrect.

Land salt, or rock salt, is mined from huge underground deposits that were formed millions of years ago when climate changes dried up large bodies of salt water. All of our salt, therefore, comes from a sea, either ancient or contemporary. But does today's ocean salt contain more minerals than mined salt? Yes, it does, if by "ocean salt" you mean the sticky, gray solids that are left behind when you evaporate every bit of water from a bucket of ocean. We'll call this crude material *ocean solids*.

Only about 78 percent of ocean solids is sodium chloride, or common salt; 99 percent of the rest is magnesium and calcium compounds. Beyond that, there are at least seventy-five other chemical elements in very tiny amounts. To get the amount of iron in a single grape, for example, you'd have to eat a quarter of a pound of ocean solids—two pounds to get the amount of phosphorus in that grape. Considering the half-ounce or so of salt that an American eats per day, ocean solids are about as valuable nutritionally as sand.

But even if it really does come from the sea, the stuff they're selling in your health food store isn't even crude ocean solids. It has been refined every bit as thoroughly as land salt has, because federal standards require that all salt sold for table use must be at least 97.5 percent sodium chloride. In practice, it's usually closer to 99 percent. One exception is a brand of sea salt imported from France, which nevertheless still has a mineral content well below that of crude ocean solids.

At a typical sea salt plant, the sun is allowed to evaporate most of the water. The solids that crystallize out, called *solar salt*, are separated from the remaining liquid, called the *bitterns*. Now, whenever a chemical compound crystallizes out of a liquid, it leaves almost all impurities behind. (Chemists use crystallization as a deliberate purification process.) The bitterns that remain behind therefore retain virtually all of the calcium, magnesium, and other "precious mineral nutrients," as the sea salt labels like to call them. In Japan, the bitterns often find their way to the table as a unique bitter condiment called *nigari*, but in the United States they're either discarded or sold to the chemical industry, where the minerals are extracted and put to various uses.

But that's not all. Next, the sea salt is washed, which removes even more calcium and magnesium because their chlorides are more soluble in water than sodium chloride is. And finally, adding insult to purity, the salt may be kiln dried using heat from—guess what?—burning coal or oil. So much for the environmental innocence of sea salt.

The product that ends up in the stores contains only about one-tenth the minerals of the original ocean solids. To get that single grape's worth of phosphorus, you'll now have to eat about twenty pounds of this stuff.

And then there's the virtually unshakable belief that sea salt is rich in iodine, "the aroma of the sea." More baloney. Certain seaweeds are indeed quite rich in iodine, but only because they concentrate it out of the water, much as mollusks extract calcium for shell-building. The seaweed story has been so widely touted that many people believe that the ocean is one big kettle of iodine. But ounce for ounce, even as unlikely a source as butter contains about twenty-four times as much iodine as the crude ocean solids. Iodized table salt, whether originally from land or sea, contains about sixty-five times as much, deliberately added at the packaging plant.

Flavor? Another fairy tale. To listen to various food gurus, sea salt tastes saltier, sharper, more delicate, more bitter, and less chemical (whatever that means) than regular table salt. The only claims that are based on a grain of truth are the ones about bitterness and saltiness. The rest are pure hokum.

The magnesium and calcium chlorides in crude ocean solids are indeed bitter. Some people have therefore been fooled into imagining that store-bought sea salt also has a bitter tang. It doesn't. The magnesium and calcium compounds never made it to the store. (The one exception, again, is that imported French salt.)

Oddly enough, however, the *saltiness* of various salts is worth discussing, even though they're all virtually pure sodium chloride. That's because different products can have different grain shapes and sizes, depending on how they were crystallized from the brine during the refining process. They range from cubes to pyramids to irregular flakes.

The most common mined table salt is in the form of tiny cubes, whereas many sea salt products, but by no means all, tend to be flaky (see p. 69). Because flakes dissolve faster than cubes, they may give you a quicker rush of saltiness when you put some on your tongue. A person who is taste-testing a flaky sea salt against a granulated land salt might attribute the slightly saltier effect to the sea, rather than to the flakiness.

Taste-testing salt is pointless anyway, no matter how carefully you do it, because people do not eat salt by putting it directly on their tongues. They add it to their food, either during cooking or at the table. In either case, the instant the salt hits wet food, it dissolves, and any conceivable effects of grain shape disappear. Moreover, when a teaspoon or so of salt is thrown into a pot of stew, any purported flavor differences would be so watered down as to be undetectable.

When you season with salt in cooking or at the table, then,

it makes absolutely no difference what kind of salt you use. The next time you hear an expert pontificating about the virtues of sea salt, take it with a grain of you-know-what. Preferably, ordinary granulated you-know-what. It's a heck of a lot cheaper.

Pepper Sí, Salt No!

I've heard it said that the quality of a restaurant's food is inversely proportional to the size of its pepper mill. Be that as it may, what about salt mills? One place that I've heard of proffers (and sells) a salt grinder for "freshly ground salt." Is it any good?

Yes, it's very good for those who manufacture the salt mills and peddle them to yuppies who must own the ultimate pseudo-gourmet accessory. For the rest of us? It's a scam. Pepper should be ground fresh, but grinding salt accomplishes nothing, but exercise.

Freshly ground black or white pepper (which are identical plant berries that have been processed differently) is indeed much superior to that dead gray powder you buy in a can. The reason is that the main flavoring chemicals in pepper are rather volatile; they evaporate gradually into the air when the dried berries are broken apart. Pepper mills, therefore, are an absolute necessity in the kitchen and on the table. They release the full flavor and aroma of the spice on demand.

But salt is a completely different story, in spite of the fact that we tend to think of salt and pepper as being related because they're eternally inseparable in recipes. And that's just what the salt-mill pushers are counting on. The fact is that it matters not a whit whether you grind your salt "fresh" or store it in the cupboard. After all, before you got it, it languished for millions of years in the salt mine without going the least bit stale.

Salt, or sodium chloride, is, of course, a mineral, not a

vegetable product. It is the only edible rock. It is made up, through and through, of sodium and chlorine atoms and nothing else. A chunk of salt is in many respects like a chunk of glass; crack it open and all you've got is several chunks, identical in every respect to the original except in size and shape. There is nothing extra inside to be released, and nothing is changed by grinding except that the chunks are now smaller than they were. You can buy salt ready-made in a variety of fine or coarse textures without having to grind it yourself. And it'll be ten or twenty times cheaper than the deliberately large chunks that they sell for use in salt mills.

You didn't ask, but . . .

Why is a saltcellar so named?

It's not a corruption of *saltseller*, in spite of the fact that it's a dish of salt on the table and could be thought of as offering, or selling salt. The *cellar* part of the word is just our anglicization of the French word *salière*, which means a dispenser of salt. Since "Please pass the cellar" would get us only a blank stare, we've tacked the redundant prefix *salt* onto our English version of *salt dispenser*.

The Blivet Lives!

My recipe tells me to dissolve two cups of sugar in one cup of water! It won't fit, will it?

Why didn't you try it?

TRY IT *Add a cup of water to two cups of sugar (or vice versa) in a saucepan and stir while heating slightly. All the sugar will dissolve.*

The concept of the *blivet*—ten pounds of stuff in a five-pound bag—has amused many generations of little boys. But dissolving sugar in a cup of water is very different from stuffing it into a one-cup sack, and one of the reasons is very simple: The sugar molecules can squeeze into empty spaces between the water molecules, so they are not really taking up any new space.

At the submicroscopic level, water isn't a densely packed pile of molecular particles, like a pail of sand grains. It's a somewhat open latticework of molecules that are held to one another end-to-end in tangled strings, rather than in a randomly oriented heap. The holes in this open latticework of molecules can accommodate a large number of dissolved particles—not only sugar molecules, but a great many others. That's one reason why water is such a good solvent—a good dissolver of many substances.

Perhaps more convincing, however, is the fact that two cups of sugar is a lot less sugar than it may seem. Sugar molecules are much heavier and bulkier than water molecules, so there won't be as many of them in a pound or in a cup. Also, the sugar is in granulated form, rather than in the form of a liquid, and the grains don't settle down into the cup as tightly as you might think. The surprising result is that a cup of sugar contains only about one twenty-fifth as many molecules as there are in a cup of water. That means that in your two-cups-of-sugar-in-one-cup-of-water solution, there is only one molecule of sugar for every twelve molecules of water. Not such a big deal, after all.

BAR BET *I can dissolve two cups of sugar in one cup of water. Don't use confectioner's or bartender's sugar; it contains cornstarch that will gum things up.*

The Sound of One Pan Sticking

Why doesn't anything stick to nonstick cookware? Isn't it strange that one material can have an aversion to everything else, no matter what? Come to think of it, what makes anything stick—or not stick—to anything else?

Quite obviously, sticking can't occur unless there are two distinct materials; there must be both a sticker and a stickee. The properties of both must be considered. But can there be such a thing as an intrinsic non-sticker, regardless of what the stickee may be?

This non-sticky question was settled in 1938, when a Du Pont chemist named Roy Plunkett came up with polytetrafluoroethylene, which was quickly and mercifully trademarked Teflon. PTFE (as we'll call it) is a remarkably unfriendly chemical compound that appears to reject the formation of lasting intimate relationships with anything and everything.

After appearing in a variety of industrial guises, such as slippery bearings that don't need oil, Teflon began to show up in the kitchen in the 1960s as a coating for frying pans that would clean up in a jiffy because they didn't get dirty in the first place. Food just wouldn't burn onto them. In today's fat-phobic society, the major virtue of nonstick pans seems to be that you can sauté in them with very little oil.

Modern variations on the nonstick theme are known by a variety of trade names, but they're all PTFE, coupled with various schemes to make it stick better to the pan, which, as you can imagine, is no small trick. (See below.)

What's really going on when one object sticks to another? It's pretty clear that there must be some sort of attraction between the two objects. The degree of stickiness depends on the strength of that attraction and how long it lasts. Glues are substances that are deliberately created to form strong, permanent attractions to as many substances as possible. But ordinary stickiness, like that of a lollipop to a child or an egg to a frying pan, is a much weaker attraction that can generally be overcome with a little physical encouragement.

If you can't unstick something with a little muscle, however, you might have to resort to chemistry. Paint thinner (mineral spirits) will usually get the chewing gum off your shoe when no amount of scraping will do the trick. So we are led to the conclusion that things may stick to one another (and become unstuck from one another) for reasons that are primarily either physical or chemical.

Why does an egg tend to stick to a stainless steel or aluminum frying pan? First of all, unless it is polished like a mirror, the surface of any metal will inevitably contain microscopic crags and crevices—not to mention tiny and not-so-tiny scratches from use—that a congealing egg white can grab onto. That's physical sticking. To minimize this kind of sticking, we use oil. It fills in the crevices and floats the egg above the crags on a thin layer of liquid. Any liquid would do, of course, but water wouldn't last long enough in a hot pan to do much good unless you used lots of it, in which case you'd have a poached egg instead of a fried one.

The surfaces of nonstick pan coatings, on the other hand, are extremely smooth on a microscopic scale. Because they have virtually no cracks, there's nothing there for food to grab onto. Of course, many plastics share this virtue, but PTFE also stands up well to high temperatures.

So much for physical, or mechanical, sticking. But the chemical causes of sticking can be more important. After all, molecules do have a tendency to form attractions to one another; that's what chemistry is all about. The atoms or molecules in a frying pan surface can form certain kinds of bonds to some of the molecules in the foods. The question now becomes, what is there about the molecules of such coatings as Teflon, SilverStone, and others that makes them inherently unreactive, or inert, toward the molecules of virtually everything else? The answer lies in the uniqueness of PTFE as a chemical compound.

PTFE is a *polymer*—a substance consisting of large numbers of identical molecules, all strung together to form huge super-molecules. PTFE's molecules are made up of only two kinds of atoms, carbon and fluorine, in a combination of four fluorine atoms for every two carbon atoms. Thousands of these six-atom molecules are bonded together into gigantically bigger molecules that look like long chains of carbon atoms with fluorine atoms bristling out like the spines on a giant wooly caterpillar.

Now, of all types of atoms, fluorine is the one that least wants to react with anything else once it is comfortably bonded to a carbon atom. PTFE's bristling fluorine spines therefore effectively constitute a suit of armor, protecting the carbon atoms against being tempted to form bonds with anything else that comes along. And that includes the molecules in an egg, a pork chop, or a muffin.

More than that, PTFE won't even let most liquids adhere to it strongly enough to wet it. (Put a few drops of water or oil on a nonstick pan and see.) If a liquid can't wet a surface, any chemicals that may be dissolved in it, no matter how powerful, just can't get a grip on the surface long enough to react with it. So no chemicals react with PTFE.

Have you detected a problem? Right. How do they get the PTFE to stick onto the frying pan in the first place? They use a variety of clever physical, rather than chemical, techniques to make the pan's surface rough enough for the PTFE coating to latch onto. These roughening techniques are the major differences between brands of nonstick cookware.

You didn't ask, but . . .

How do those no-stick cooking sprays work for low-fat frying and baking?

They are nothing but cooking oil dissolved in alcohol, in a handy aerosol can. The idea is that instead of pouring a random amount of oil into your skillet, you give it a quick spritz from the can. The alcohol evaporates and the oil coats the pan. You'll still be cooking on a "no-stick" layer of oil, but it's a very thin—and low-Calorie—one.

A tablespoon of butter or margarine contains about eleven grams of fat and one hundred Calories. On the other hand, the labels of the cooking sprays brag about containing "only two Calories per serving." A "serving" is defined as a one-third-of-a-second spray, which, they oddly advise, is just long enough to cover one-third of a ten-inch skillet. But even if your trigger finger isn't as finely trained as Billy the Kid's, or if you throw caution to the winds and cover the entire pan, you can still manage to get by with very little fat.

By the way, if you're a belt-and-suspenders type, spritz a little no-stick spray onto your nonstick frying pan. The food will brown better than it would without the fat.

Waiter, There's Salt in My Jam

How does sugar preserve fruits and berries, such as in the making of jams and preserves? I suppose the sugar must kill germs somehow, but I've never thought of sugar as a germicide. And if sugar is lethal to bacteria, why isn't it at least a little harmful to us?

There's nothing unique about the use of sugar to preserve foods. In principle, you could make your strawberry jam with salt instead of sugar and it would keep just as long. Much longer, in fact, because nobody would go near it after the first taste. Salt has been used for thousands of years, however, to preserve fish and meats. The wonderful cured salmon called *gravlax* is usually made with a mixture of salt and sugar.

Although sugar and salt work quite well in killing or deactivating microorganisms to keep foods from spoiling, they function only when they are in very concentrated form. You can't sterilize foods just by sprinkling these familiar kitchen chemicals on them. But if you use *enough* sugar or salt, so that when it dissolves in the foods' juices it makes a solution of at least 20 or 25 percent, then most bacteria, yeasts, and molds simply can't survive. And no, it's not because they die from diabetes or high blood pressure.

What happens is that the sugar or salt solution sucks most of the water out of the little buggers—dehydrates them—so they just shrivel up and either die or become inactive. Practically nothing can live indefinitely without water, and these microscopic, one-celled organisms are no exception.

How can a solution of salt or sugar pull water out of an object? The answer is by *osmosis*, that seemingly all-purpose word that people throw around to refer to any mysterious kind of seepage. ("I never really studied Urdu, but my parents spoke it and I guess I got it by osmosis.")

Osmosis is actually only one particular kind of seepage. It's the seepage of water through a thin membrane, and it occurs whenever there happen to be two solutions of different concentrations (strengths) on opposite sides of the membrane. The membrane must be *semipermeable*. It must allow water molecules to seep through, but not other molecules. Most of the thin, sheet-like membranes that separate organs from one another in plants and animals are semipermeable. In our own bodies, that includes the walls of our red blood cells and the walls of our capillaries.

In osmosis, there is a net transfer of water molecules through the membrane from one of the solutions to the other, but not in the reverse direction. In a sense, the membrane functions as a one-way street for water molecules. The direction of the traffic depends on the relative concentrations, or strengths, of the two solutions. The water will flow from the less concentrated solution to the more concentrated one. Let's see how that happens in the case of those villainous bacteria on your strawberries.

A bacterium is essentially a tiny blob of jelly-like protoplasm encased within a cell wall that functions as a semipermeable membrane. The bacterium's protoplasm is water with various kinds of stuff dissolved in it—proteins and many other chemicals that are terribly important to the bacteria but of little concern to us at the moment.

Now, let's deluge this blob-within-a-membrane with a flood of very salty or sugary water. Suddenly the concentration of dissolved matter outside the cell is higher than it is inside. That means that there are relatively fewer freely-moving water molecules in the outside solution, because they are hindered by the dissolved substances.

We thus have an unbalanced situation in which there are different concentrations of free water molecules on the two sides of a thin, water-penetrable membrane. Now, Mother Nature just hates imbalances, and she always tries to even things up if she can. In this case, the balance can be restored if some of the free water molecules on the inside migrate through the membrane to the outside. And that's just what happens.

Osmosis behaves almost as if there were some kind of pressure forcing water through a membrane from the low-concentration side to the high-concentration side. Scientists actually do talk in terms of an *osmotic pressure*, and they deal with it pretty much in the same way as they deal with gas pressures.

For our hapless bacterium, the net result is that water will have been sucked out of it, whereupon it promptly bites the dust. At the very least, it is so weakened that it is incapable of reproducing. ("Not tonight, dear, I'm dehydrated.") In either case, the threat to our health has been eradicated.

For the same reason, shipwrecked people, when stranded in a lifeboat or on a raft at sea, can't drink any of the "water, water everywhere." Perversely, drinking this water could dehydrate them fatally.

The same fate is likely to befall a fresh-water fish when put into salt water. Osmosis will draw water out of the fish's cells and into the saltier sea, and the fish can die of dehydration—a rather ironic mode of demise for a fish.

Wimpy's Revenge

If I had a strong enough magnet, could I lift spinach?

Not unless it was in a steel can, which is perversely known as a tin can, even when it's an aluminum can. The much-touted iron in spinach just isn't in a form that's attractive to a magnet.

Iron is magnetic—attracted to magnets—when it is in the metallic form (see p. 231), but not when it is chemically combined with other elements. The metallic iron in your steel refrigerator door can attract a whole conglomeration of silly looking magnetic doodads, but iron in the chemical form of rust, for example, is nonmagnetic. It's the same in the case of spinach: The iron in the spinach (fortunately) isn't in the form of little pieces of metal; it's in the form of complex chemical compounds that just aren't magnetic.

But why does everybody think of spinach when they think of iron-rich foods? Probably the major reason is Popeye, that irrepressible cartoon character who for more than sixty years has been demonstrating to the world that a combination of virtue, spinach, and stupidity will triumph in the end.

Actually, there's nothing unique about the iron in spinach. Many green vegetables and other foods of sundry colors also contain substantial amounts of iron. Ironically (accidental, so help me), a hamburger contains about the same amount of iron as an equal weight of spinach. So how come spinach made Popeye powerful, while hamburgers made Wimpy wimpy?

Popeye was simply helping the mothers of America to get their kids to eat their vegetables—especially spinach, which tastes lousy to most kids because of the sour oxalic acid that it contains. (Just try to get a child to eat rhubarb, which packs a mouth-puckering load of oxalic acid.) If Daddy didn't happen to be an adequately muscular role model ("Don't you want to grow up big and strong like . . . ?"), Mom could always use Popeye as an indisputable surrogate.

So much for mineral nutrition. But what about Popeye's fabled strength? Why didn't he guzzle cans of squash or turnips instead of spinach? How did cartoonist Elzie C.

Segar, Popeye's creator, ever decide that spinach should be his sailor man's ticket to brawniness?

It's that legendary iron again. People who don't have enough iron in their blood are often pale and feeble. The adjective *anemic* has actually come to mean weak or lethargic. Of course, that doesn't mean that eating more iron will make you stronger if you're not anemic. But since when has a cartoon character ever been deterred by logic?

3

In The Garage

Your car is rusting away before your eyes; it won't start; your tires are flat; and you've just skidded on your icy driveway, hitting a tree branch and shattering your shatterproof windshield. Wouldn't it be comforting to understand all the science behind these events? Well, maybe after you've calmed down.

Here's where we look at some of the fascinating phenomena that have grown out of our love affair with the infernal combustion engine.

A Shocking Dilemma

In cold weather, my car's battery acts half dead. In real cold weather, it won't even start the car. Yet I've been told to keep my flashlight batteries in the refrigerator to keep them lively. Why is cold good for flashlight batteries but bad for car batteries?

Nobody's telling you to try to use the flashlight batteries when they're cold. They would be just as sluggish as your car battery. Cold inhibits both kinds. They must be somewhere around room temperature if you are to get the intended amount of "juice" out of them.

Batteries produce their electricity—streams of electrons—by a chemical reaction (see p. 39), and all chemical reactions go more slowly at lower temperatures (see p. 156). Cool a battery much below normal room temperature and the number of electrons it can put out per second (Techspeak: the amount of *current* it can deliver) is severely limited, whether it's a battery in your car or in your flashlight. Cold batteries in your Walkman will turn your *allegro vivace* into *lento*. And by the way, don't put them in until they've warmed up, or else condensation of moisture on the cold surfaces will give you water music, and I don't mean Handel's.

It is only the battery's ability to deliver current—robust streams of electrons on demand—that is inhibited by cold temperatures. Cold has virtually no effect on the force with which the battery sends out the electrons (Techspeak: the *voltage*).

Another thing: Batteries leak a bit of electricity even when they're not hooked up, that is, even when they're not delivering intentional amounts of electricity. This eats into their limited supply of chemicals. If you keep them cold, you're slowing down even this small amount of chemical reaction and preserving the power for when you really need it. But today's alkaline batteries have such a long shelf life that refrigeration will hardly make a difference.

In automobile batteries, which contain a liquid (sulfuric acid), there is another cold-restricted factor. When the battery is delivering current, certain atoms (actually, they're ions, but I won't tell anybody if you don't) must migrate, or swim through the acid from the positive internal pole to the negative pole and vice versa. At cold temperatures they are substantially slowed down, so the battery's ability to deliver current is also inhibited.

Some old-time garage mechanics will swear to you that if they leave a car battery on the concrete floor for a long time instead of on a shelf, the concrete "sucks the electricity out of it." What's going on, of course, is that the floor is cold and sucks the heat out of it.

What you *really* have to watch out for is the mechanic who sucks the money right out of your wallet.

A Shattering Experience

For obvious reasons, automobile windshields are made so that when they are smashed, the pieces don't fly all over the place. But why do they break into so many tiny pieces instead of into a few big ones? How do they get glass to break up that way?

Preventing the scatter of fragments is relatively easy. The windshield is actually a sandwich, with glass "bread" and an elastic plastic "ham" that can be indented without cracking. When the bowling ball hits the windshield, most of the pieces of glass remain bonded to the plastic instead of flying around loose. But why it breaks into a million small fragments instead of the small number of pieces that you'd get by smashing a sheet of ordinary glass is another question. It has to do with how the glass is tempered—pretreated—to make it stronger.

Windshields, of course, do have to be stronger than ordinary glass. To make a material stronger, engineers often resort to *prestressing* it—subjecting it to certain forces. And that's what they do to the windshield glass.

While the glass is still at a high temperature after being formed into shape, the surfaces—and only the surfaces—are instantaneously chilled. This locks in the molecular structure of high-temperature glass, which has a more expanded structure than room-temperature glass. When the whole sheet is then allowed to cool slowly down to room temperature, it retains the high-temperature structure frozen into its skin, while the interior shrinks down to the tighter room-temperature structure. Thus, a combination of opposing tension (pulling) and compression (pressing) forces has been locked into the glass—a sort of pent-up push-me-pull-you contest that strengthens the entire structure.

This pent-up energy is released the instant the glass becomes flawed or cracked anywhere. Utilizing this energy, the fracture quickly spreads like a chain reaction over the stressed surface. Because every part of the surface is stressed, the cracks and breaks erupt equally all over it, resulting in the familiar gravel-like pattern of a million pieces.

You didn't ask, but . . .

How do they prestress concrete?

They play hard rock music while it's being poured.

Sorry.

The strength of prestressed concrete doesn't come from heat tempering, as in the case of windshield glass. Prestressed concrete is concrete containing steel cables that have been subjected to tension—that is, that have been stressed by pulling on them lengthwise, before the concrete hardens. The cables then want to contract as if they were incredibly stiff rubber bands, but since they can't, they keep the hardened concrete under constant compression. In a sense, some of the stress energy from the original tension-pull has been locked into the structure in the form of compression, and that makes it stronger, because concrete is very strong under compression, but under tension it isn't worth its weight in taffy.

Busting the Rust Crust

Everything I have is rusting away. Well, not really, but it seems I'm always fighting off rust by oiling, scraping, and painting everything I own, from tools to lawn mowers and porch railings. I won't even mention automobiles. Maybe if I knew more about what causes rust in the first place, I could head it off. Any help?

Iron plus oxygen plus water equals rust. That's it. When all three are present, rust will inevitably occur. But if any member of this unholy triumvirate is missing, there can be no rust.

Fortunately for us living creatures, but unfortunately for our garden tools and automobiles, oxygen and water vapor are present everywhere in the atmosphere. And fortunately or unfortunately, the entire center of our planet, a core that measures some four thousand miles in diameter, is almost 90 percent iron. Even the sun and the stars contain iron.

Here on the surface of the Earth, from whence we dig our minerals, iron is the most abundant of the eighty-eight known metallic elements. It is therefore the cheapest of all metals and the most widely used, whether in the form of wrought iron, steel (iron with carbon in it), or any one of dozens of alloys.

You just can't get away from iron, oxygen, and water, so it's no wonder you have a problem. But you're not alone. The rusting of iron has plagued mankind since prehistoric times.

The main villain is oxygen. In a process called *oxidation*, it reacts with most metals to form *oxides*, and rust is a form of iron oxide. (In chemical circles, it goes by the name of *hydrated ferric oxide*.) Under the right conditions, oxygen will also react with aluminum, chromium, copper, lead, magnesium, mercury, nickel, platinum, silver, tin, uranium, and zinc, among many other metals. In fact, among all the metals that you're probably familiar with, only gold is completely immune to its attack. That fact, plus gold's scarcity and unique color, is what makes it so highly prized.

(Incidentally, those jewelry-cleaning products that claim to "remove tarnish" from your gold jewelry are a fraud. Gold doesn't tarnish. Plain soap and water will remove any dirt.)

Oxidation doesn't corrode, deface, and destroy any other

common metal the way it does iron. That's because most other metals have some sort of saving grace that keeps the oxygen from chewing them up. For example, oxygen reacts very readily with aluminum, but it happens that the first thin layer of oxide on the metal's surface is so tough and airtight that it seals off the rest of the metal from further attack. Other metals, such as copper, for example, react so slowly that all they do is darken a bit (but see page 148). The coating of oxide then protects the rest of the metal from severe corrosion.

When oxygen and water attack iron, however, the reddish-brown oxide doesn't stick. As you know from sad experience, it tends to flake off and crumble away, uncovering ever-fresher surfaces of metal for the air and moisture to ravage. Iron oxide's molecular arrangement just happens to make it a weak and crumbly material, and there is nothing we can do about it. There are products on the market, however, that can convert the structure of rust into a tough, adherent coating. Check your hardware store.

The only lines of home defense against rusting, then, involve keeping the iron away from prolonged contact with moisture or oxygen. Never put your tools away wet. And anything that will fit into an airtight plastic bag can rust only as long as the limited amounts of oxygen and water vapor in the bag hold out. Sorry, but that's about all you can do, short of painting.

TRY IT

Even when immersed in water, iron won't rust if there is no oxygen present. Boil some water vigorously for several minutes to get most of the dissolved air out of it and then let it stand overnight in a glass jar. Fill a similar jar with fresh tap water. Drop an iron nail into each jar and wait a couple of days. The nail in the boiled water will rust a lot less than the one in the tap water. (Boiling can't remove every bit of oxygen.)

You didn't ask, but . . .

Why does salt make cars rust faster, whether it is present in the air near the ocean or whether it is being used on highways during freezing winters?

Rusting takes place through a juxtaposition of iron and oxygen that actually constitutes a miniature electric battery, on the atomic scale. That is, the oxygen molecules are taking electrons away from the iron atoms, and that is exactly what goes on inside a battery: electrons being snatched from one substance by another (see p. 39). Anything that helps electrons to go from the iron atoms to the oxygen molecules will help this process along.

Salt helps because when salt dissolves in water it makes a solution that is a good conductor of electrons. Therefore, salt helps iron to rust by helping to deliver the iron atoms' electrons to the voracious oxygen molecules.

NITPICKER'S CORNER:

In the rather complex atom-by-atom mechanism of rusting, salt also helps to conduct charged iron atoms (ions) to where they need to go. Moreover, the chloride in the salt, which is sodium chloride, has a separate effect on the iron. But that's all a bit more than we want to get into. Trust me. Just don't drive your car in salt water.

Help! My Antifreeze Froze!

Expecting an unusually cold winter, I drained my car's cooling system and put straight antifreeze in it instead of the usual fifty-fifty mixture with water. Now a mechanic tells me that straight antifreeze freezes at a warmer temperature than 50 percent antifreeze does. How is that possible?

Strange as it may sound, your mechanic is correct. A fifty-fifty mixture of ethylene glycol and water won't freeze until the temperature gets down to about 34 degrees below zero Fahrenheit (–37 degrees Celsius), while pure antifreeze will freeze at about 11 degrees *above* zero (–12 degrees Celsius). Let's see what's going on here.

It happens that mixing almost anything at all into water will lower the freezing point below water's normal value of 32 degrees Fahrenheit (0 degrees Celsius). In principle, you could add salt, sugar, maple syrup, or battery acid to your engine coolant and they'd all work to some extent, but for obvious reasons they're not recommended.

In the earliest days of automobiles, they did occasionally use sugar and honey as antifreeze. Later, alcohol became popular, but it boils off too soon. These days we use a colorless liquid chemical called ethylene glycol, which doesn't boil off. Commercial antifreeze also contains rust inhibitors and a vivid dye to help locate leaks in the cooling system and, not incidentally, to make it look technologically sophisticated.

The freeze-protecting power of dissolved substances in water has to do with a fundamental difference between the arrangements of the molecules in liquids (such as water) and in solids (such as ice).

In water, as in all liquids, the molecules are slithering around freely like a mass of oiled bodies at an orgy. They're loosely attracted to one another, but they're not connected in fixed positions, as they are in most solids. That's why you can pour a liquid, but not a solid.

In order for liquid water to freeze, then, the molecules must slow down and settle into the highly proper, rigid positions that they must occupy in a crystal of ice. If given enough time to find these positions—that is, if the molecules are slowed down gradually enough by gradual cooling—water is capable of forming rather large chunks of ice. And that's exactly what we're afraid of, because when water freezes, it expands (see p. 202) and the consequent pressure can crack the cooling passages in the engine block.

Alien molecules in the water, such as ethylene glycol, for example, throw a monkey wrench into this freezing process in two ways. First, by simply cluttering up the place, they interfere with the water molecules' ability to fall into those precise locations that are needed to form a crystal of solid ice. It's as if a military drill team were trying to fall into formation while a mob of civilians is running around on the field. By getting in the way, the alien molecules prevent the ice crystals from growing to be as large and uniform as they would like. Even if the water does freeze, then, the result will be a slush of small ice crystals, rather than a single, rock-hard, engine-cracking iceberg.

But the main effect that extraneous molecules have on freezing water is that they keep the water from freezing until a lower-than-normal temperature. What happens is that the ethylene glycol molecules are "diluting" the water, thereby reducing the number of water molecules that can congregate in any one spot to form an ice crystal. Because of that, we have to slow down even more of the water molecules by continuing to lower the temperature, in order to get enough of them to fall into place together as an ice crystal.

Why then, will pure ethylene glycol freeze at a higher temperature than a 50 percent mixture with water? It freezes more quickly because ethylene glycol's molecules are interfered with by water molecules, just as water's molecules are interfered with by ethylene glycol molecules. It works both ways. The water lowers ethylene glycol's freezing point, even as the ethylene glycol lowers the water's freezing point. So ethylene glycol mixed with water won't freeze as easily as it does when it is pure.

Yes, it can be said that water keeps antifreeze from freezing.

You didn't ask, but . . .

The label on the antifreeze container says that it not only keeps the coolant from freezing, it also keeps it from boiling. How is boiling related to freezing?

By interfering with water's molecules, dissolved substances not only lower its freezing point, they also raise its boiling point by making it harder for the water molecules to fly off into the air (see p. 48). With ethylene glycol dissolved in it, your coolant water has to get to a higher temperature than usual before it will boil. A mixture of 50 percent ethylene glycol in water won't boil until 226 degrees Fahrenheit (108 degrees Celsius). That's less of an advantage than it used to be, however, because today's cooling systems are pressurized, and at elevated pressures the boiling points of both water and ethylene glycol are already higher than they would be at atmospheric pressure (see p. 208).

BAR BET In a car's cooling system, straight antifreeze will freeze sooner than a mixture of antifreeze and water. Water keeps antifreeze from freezing.

Car-Skiing: A One-Time-Only Sport

I live in a cold climate, and my house has a steeply sloped driveway. When the driveway is icy, I sprinkle sand on it to improve the traction of my tires. But the last time I tried it (and it will be the last time), the sand didn't work. It acted like so many tiny ball bearings under my tires, with highly unpleasant consequences. Why didn't the sand give my car traction? (It almost put me in traction.)

It was an extremely cold day, wasn't it? Below zero Fahrenheit, perhaps? That was the problem. Sand won't work when it is too cold.

In order to improve traction, the sand grains must become partially embedded into the ice, making tiny bumps in what had been a smooth surface—in effect, making "sandpaper" out of the ice. It is the pressure of the car on the sand that accomplishes this. When the tire presses a sand grain against the ice, a bit of the ice melts beneath the grain, and it sinks in. The water then refreezes around the grain.

The ice melts under this pressure because ice is the bigger-volume form of water, and when pressed upon it reverts to its smaller-volume form: liquid water (see p. 211). Without this pressure-melting effect, the sand could not embed itself into the ice.

The problem is that the colder the ice is, the more pressure is needed to melt it, because the water molecules in the ice crystal are more rigidly fixed in place and cannot easily be persuaded to move around loosely, as the molecules of a liquid do. Even though a car applies a lot of pressure to a grain of sand, it may not be enough to melt the ice in below-zero weather.

You might do better on foot. A rubber tire isn't the greatest pressure-applying device because of its elasticity. The soles of your shoes are probably harder, and even though I presume that you weigh less than one-fourth as much as your car (one wheel's worth), you may still be applying more pressure to the sand grains—more pounds per square inch of grain—than the car does, and the sand will embed itself by the melting mechanism.

The Salt Man Cometh

When there's ice on my driveway, I throw salt on it and the ice melts. But how can anything melt ice without heat? They say it's because salt lowers the freezing point of water, but what can that possibly mean to the ice? It's already frozen.

Contrary to what everybody says, the ice on your driveway doesn't *melt*, any more than sugar *melts* in coffee or tea. People often confuse melting with dissolving. ("I don't need an umbrella; I won't melt in the rain.") But melting, as you have already noted, requires heat. You can certainly melt ice or sugar (see p. 63) by heating them, but that's not what the salt does to the ice. The salt *dissolves* the ice.

People use the word *melting* for the salt-on-ice phenomenon only because they see ice disappearing and a liquid—salt water—remaining. And *melt* just happens to be the word that our ancestors invented for "ice go 'way, water come." Science teachers and textbooks, however, should know better than to fall into that linguistic trap.

Along with many others, you were probably taught in school that "salt lowers the freezing point of water." But that's not literally true either. Throwing salt onto your driveway can't possibly change the freezing point of water—the temperature at which good old aitch-two-oh is accustomed to freezing or melting. That temperature—it's the same for melting as for freezing (see p. 63)—is 32 degrees Fahrenheit or 0 degrees Celsius. It always has been, and it always will be. What the textbooks and teachers *should* be saying is that *salt water* freezes at a lower temperature than *pure water* does (see p. 92). That's quite a different statement.

On your driveway, the salt first makes salt water out of the ice, and then the resulting salt water remains unfrozen because *its* freezing point—not *water's* freezing point—is indeed below the temperature of the air. A fine distinction, perhaps, but critical to understanding what's going on.

First, how does the salt make salt water out of the ice? It happens that sodium and chlorine atoms (actually, sodium and chlorine ions, but we won't quibble), which make up the sodium chloride, or salt, have a strong affinity for water molecules. (Manufacturers have to add an anticaking agent to keep table salt from gumming up in the shaker because of moisture absorbed from the air.) When a crystal of salt lands on an ice surface, the salt's sodium and chlorine atoms pull some water molecules out of the surface. They then proceed to dissolve in that water to form a tiny puddle of salt water around the crystal. The puddle of salt water doesn't freeze because *its* freezing point is lower than the temperature of the air.

The sodium and chlorine atoms that are now dissolved in

the salt water keep nipping away at the ice surface like piranhas going after a meatball in a punch bowl. As the process continues, more ice continues to dissolve in the salt water, making more and more salt water. Eventually, either all the ice runs out, or the puddle of salt water becomes so dilute that its freezing point is no longer below the air's temperature, and it will freeze. But salt water freezes only into slush, rather than into hard ice. In either case, your ice-destroying mission has been accomplished.

BAR BET *Salt doesn't melt ice.*

The Duck's Back Caper

Why won't oil and water mix?

Ordinarily, water is the best mixer in the world, and I don't mean just with Scotch. It mixes with, it associates intimately with, it even welcomes into its very bosom—that is, it *dissolves*—more substances than any other liquid. That's why it is sometimes called the universal solvent.

But there is one family of substances that water abhors and will invariably shun: oils. Water won't even snuggle up close enough to a drop of oil to wet it, much less to dissolve it. Water rolls off a duck's back because the duck's feathers are oily, and they don't even get wet when the duck goes diving. But you knew that.

Like guests at a social gathering, molecules must have at least something in common in order to mix successfully. Quite literally, the molecules of water and oil have practically nothing in common. Water, as you may know, consists of small, three-atom molecules: two hydrogen atoms and one oxygen atom. Oils, on the other hand, are made of big molecules consisting of many carbon and hydrogen atoms, with no oxygen at all. No matter how intimate the gathering, it is not very likely that the twain are going to meet and form an alliance.

What is there about oils that makes them outcasts in the big, wide, wonderful world of water, the most plentiful liquid on earth? Once we see why water is such a powerful solvent (dissolver) for so many other substances, we'll see that oils simply don't have what it takes to be dissolved into water.

In pure water, as in any liquid, the molecules are being held together by some kind of mutual attraction. If they weren't, they would go flying off into the air and the liquid wouldn't be a liquid any more; it would be a gas. The attractions between the molecules in water are rather special. They come from the fact that water molecules are *polar*: They are like tiny bar magnets, but instead of having north and south *magnetic* poles at their opposite ends, they have positive and negative *electric* poles—that is, positive and negative electric charges (see p. 66).

Now, if you think of a glass of water as a glass full of tiny magnets all stuck together, you can see that they would have precious little interest in associating with any substance whose molecules weren't also magnets. Magnets are attracted to nothing but other magnets. Yes, a magnet is attracted to a piece of ordinary iron, but inside that iron are zillions of tiny magnets (see p. 231).

Only if a substance contains atoms or molecules with electric poles will water be attracted to it, first by wetting it and eventually by enveloping it and dissolving it. Lots of substances fit this bill and will mix with water, but oils absolutely don't and won't because there is nothing at all polar—no electric poles—in those big, long molecules of oil. So there is nothing that might be attractive to a water molecule.

Dissolving is the most intimate possible kind of mixing. The molecules of one substance mix, one-by-one, in amongst the individual molecules of the other. Where dissolving is concerned, then, we might conclude that only birds of a feather are likely to want to flock so close together. Being less poetic, chemists prefer to say that "like dissolves like," meaning that only substances having molecules similar to those of water are likely to mix with water. And ditto for oil-like molecules and oil.

Generalizing even further, we can expect that a given substance, if it dissolves in anything at all, will dissolve in either oil or water, but not both. And that expectation is generally borne out. Salt and sugar (see below) dissolve in water; gasoline, greases, and waxes dissolve in oils. But never the other way around.

NITPICKER'S CORNER:

Besides the attraction of polar molecules, or "electric magnets," for one another, there's another important kind of attraction between water molecules. It's called *hydrogen bonding*. Without going into detail, let's just say that it can come about when molecules have an oxygen atom plus a hydrogen atom—a so-called hydroxy group, *OH*—at one end. Water molecules fit this prescription precisely, and they stick together by hydrogen bonding as well as by polar attraction.

On the "like dissolves like" theory, other substances that are good setups for hydrogen bonding should also be likely to dissolve in water. And they are. Sugar (sucrose) is a familiar example. It dissolves in water, not because its molecules are "electric magnets," but because they contain the water-like hydroxy group and therefore hydrogen-bond to water molecules. The sucrose molecule actually contains eight hydroxy groups.

If oil molecules aren't polar, and if they don't form hydrogen bonds, then what holds them to each other? It's a totally different kind of molecule-to-molecule attraction called a *van der Waals attraction*, about which we needn't bother our heads. (But see p. 102 if you must.) Suffice it to say that these attractions are just as alien to water molecules as electric poles are to oil molecules. The revulsion of water for oil, then, is quite mutual.

Many a Slip

Why is oil so good at lubricating things?

Obviously, because it's so slippery. But what makes a substance slippery?

All liquids are slippery to some degree. A wet floor or highway—wet with water—is a well-recognized hazard that keeps many lawyers in expensive clothing. But water isn't much use as a lubricant in our engines and other machinery because it is really not all that slippery and it evaporates away.

Oil is much slipperier than water because its molecules (you knew it would be the molecules, didn't you?) can slide past each other more easily than water's molecules can. And because a liquid is nothing more than a pile of molecules, when the molecules slide *you* slide. You wouldn't be surprised to slip on a pile of ball bearings, would you?

Water molecules don't slide around as easily as oil molecules do because they have a significant amount of stickiness—relatively strong attractions to each other (see p. 98). Water's particular type of molecule-to-molecule attractions arise predominantly in molecules that contain oxygen atoms, as water molecules certainly do: Oxygen is the O in H_2O.

But oil molecules, that is, the molecules of the *hydrocarbons* that make up that gooey, black chemical mishmash called petroleum, are made up of nothing but hydrogen and carbon atoms. No oxygen atoms at all. They therefore do not stick together very well and can slide easily over one another. Hence, they are good lubricants.

NITPICKER'S CORNER:

Oil molecules have to come up with some other way of sticking together, because if they didn't stick together at all, they would go flying off into the air as a vapor and all the machinery of civilization would grind to a screeching, smoking halt.

Oil molecules stick together by what chemists call *van der Waals attractions*. They explain these forces by waving their arms around a lot and muttering about electron clouds. The story goes that when a bunch of atoms cluster together to make a molecule, they pool their electrons to form a big, squishy cloud of electrons that swarms around the whole molecule like a horde of gnats around a cluster of grapes. So when two molecules come together, the first thing they see is one another's electron clouds. Gnats meeting gnats.

So far, so good. Nobody argues with that picture, which has served chemists extremely well in explaining how molecules interact with one another. But now for the weird part. In spite of the fact that all the electrons in these clouds have the same charge (negative) and should therefore repel one another, they supposedly somehow *attract* each other and hold the molecules together. That's what Professor van der Waals said, and he got the Nobel Prize in 1910 for saying it. So go argue.

Anyway, these van der Waals attractions do hold oil molecules together—especially the bigger oil molecules that have large electron clouds—strongly enough so they don't evaporate away. But because they are joined together only by the squishiness and mushiness of electron clouds, the molecules can still slide easily past each other.

Pumping Irony

How come I can pump my bicycle tires up to sixty pounds in no time, but I have to knock myself out with the bicycle

pump just to add a couple of pounds to my car's tire, which is only inflated to thirty pounds?

It's not just the pressure you're fighting against; you also have to consider the volume. It takes many more strokes of the pump to add a "pound of air" to the car tire than to the bike tire.

What people loosely call a "pound of air" is not an amount of air, like a pound of butter; it's a *pressure*: pounds of force per square inch, generally abbreviated psi. That force is the cumulative effect of the zillions of air molecules in the tire, which are continually bombarding every square inch of the inner walls. The more air molecules you force into a tire, the more bombardment there will be and the higher the pressure will be. That's why adding more air increases the pressure.

As you've surmised, it should be harder to force air into a 60-psi tire than into a 30-psi tire. That's because the air molecules inside a tire are also bombarding the valve opening, making it harder to force more molecules through. So each stroke of your pump does indeed require more effort to overcome the bike's 60 psi of pressure than to overcome the car's 30 psi. You have to use twice as much force on the pump handle to push air into the bike tire.

Then why is it more work to pump up the car tire?

A typical car tire contains about six or eight times as much air space as a typical bike tire. In order to have the same pressure—the same rate of molecule bombardment per square inch—in both tires, there would have to be six or eight times as many air molecules present in the car tire. Therefore, to increase the car tire's pressure by each psi, you have to pump in six or eight times as much air—using six or eight times as many strokes—as you do to get a psi of pressure into the bike tire. Even though each stroke takes half the effort, you're still working more than three times as hard.

Why Inflation Is Heating Up

When I inflate my bike's tire with a bicycle pump, the tire's valve stem gets hot. I assume that it's from the friction of all that air squeezing through the narrow valve. But when I fill the same tire at the gas station, the valve doesn't get hot. What gives?

It can't be friction, because approximately the same amount of air is being forced through the valve in both cases. The answer is that when air (or any gas) is compressed—when it is forced into a smaller space—it gets hot.

When you use your hand pump, you're compressing the air in the pump, but when you use the gas station's air, you're using air that has already been compressed. The gas station's air did indeed get hot when it was originally compressed into the storage tank. But by the time you show up with your sad-looking tires, the air has had lots of time to cool off. All you are doing is bleeding off some of that stored-up air. No compression is going on, so there is no heat.

Why does compressing a gas make it hot?

Well, gas molecules are free spirits; they are flying around freely, as far apart from one another as they can get, within their confines. To force them closer together—to compress them into the confines of a tire, for example—you have to oppose their outward-flying proclivities with some inward-pushing force. When you use your pump, the sweat on your brow tells you that you are indeed putting some of your own muscular energy into that gas.

But what do the molecules do with that energy? Unable to fly so far afield anymore, they use the energy you've given them to fly faster. And faster-moving molecules are hotter molecules; heat is nothing but fast-moving particles (see p. 236). Thus, your muscular energy goes into heating up the gas in the tire.

You didn't ask, but. . . .

If compressing air makes it hotter, does air get cooler when it expands?

Definitely. And that is indeed what is happening back in the gas station's compressed-air tank as you allow some of its stored-up compressed air to expand into the outer world.

Why does expansion cool a gas? Well, if a collection of flying gas molecules is suddenly allowed to expand into a bigger space, the molecules have to push their way out against whatever happens to be occupying that space—usually, the atmosphere. Doing that uses up some of the gas's energy, and the gas molecules then move more slowly. (If the gas is expanding into a vacuum, all bets are off.) A gas whose molecules are moving more slowly is a gas that has a cooler temperature.

TRY IT The next time you fly on a humid day, watch the airplane's wing during takeoff, which is the time of maximum lift. You may see a layer of fog just

above the wing's top surface. This is an example of expansion cooling. The air going over the top of the wing is expanded, compared with the air underneath the wing. (Bernoulli's principle and all that; ask any pilot.) The expanded wing-top air may be cooled enough to condense water vapor out of the air, making a stream of visible fog.

A Pair of Highly Extinctive Gases

Carbon monoxide and carbon dioxide: what's the difference? I gather that monoxide means one "oxide" (whatever that is) and dioxide means two of them. That's all right with me, but are they both poisonous? What is their connection with automobile exhausts, kerosene heaters, and cigarette smoke?

They're both dangerous gases, but in very different ways.

Small amounts of carbon dioxide are normally present in the atmosphere (see p. 157). It gets there from volcanos, from the decomposition of plant and animal matter, from the burning of coal and petroleum, and from the opening of cans of beer, which, however, is not the primary source in spite of the way it appears in television commercials. Nevertheless, eleven billion pounds of carbon dioxide are produced annually in the United States alone, and much of it is destined to be burped into the atmosphere via the eight billion cases of carbonated soft drinks and 180 million barrels of beer that Americans guzzle each year.

Obviously, carbon dioxide can't be toxic in itself. The only real problem is that it doesn't support burning or breathing (see p. 133), and if given the opportunity, it will extinguish both fires and people. Because carbon dioxide is heavier than air, it will spill down to the lowest level and hang around like an invisible blanket, replacing the air and suffocating anything it covers. That's what happened in Cameroon, Africa, in 1986 when Lake Nios belched an enormous, six-hundred-ton bubble of volcanic carbon dioxide gas that spread out over the countryside and suffocated more than seventeen hundred people and innumerable animals.

TRY IT *Light a votive candle—a candle in a small glass cup. Don't bother to pray. Now make some carbon dioxide by pouring a little vinegar onto a few teaspoons of baking soda in a tall drinking glass. As the carbon dioxide bubbles up and fills the glass, pour it over the votive candle as if you were pouring an invisible liquid. (Be careful not to pour any of the real liquid.) The candle will go out, drowned beneath a sea of unseen gas.*

Carbon monoxide, on the other hand, is a real villain, even in tiny amounts. When breathed, it goes straight from the lungs into the blood stream, where it reacts vigorously with the hemoglobin, preventing it from doing its vital job of carrying oxygen to the cells. Oxygen deprivation ultimately leads to a condition known as death. Carbon monox-

ide is the principal cause of poisoning fatalities in the United States.

Whenever carbon-containing substances burn in air—from the gasoline in a car to the kerosene in a heater to the tobacco in a cigarette—carbon monoxide is formed to some extent. If they had an unlimited supply of air, these fuels would burn completely, all the way to carbon *di*oxide—two oxygen atoms for each carbon atom. But there is always a practical limit to how fast the oxygen can feed itself into the conflagration. So invariably, some of the carbon atoms will manage to latch onto only one oxygen atom instead of two. Result: *mon*oxide instead of *di*oxide.

Automobile engines spew out about 150 million tons of carbon monoxide in the United States each year. In a traffic jam, the carbon monoxide level in the air can build up to sickening (fatigue, headache, nausea), if not dangerous, levels. Kerosene heaters, gas space and water heaters, gas furnaces, gas ranges and ovens, gas dryers, wood stoves, charcoal grills, and cigarettes all produce carbon monoxide, and all must be vented to the outdoors or used in a well-ventilated environment.

So don't smoke and drive. Especially indoors, when the kerosene heater is on.

A Thousand Pounds of Pigeon Sweat

At a truck stop, I watched a trailer-truck driver banging fiercely on the sides of his trailer with a baseball bat. When I asked him what he was doing, he explained, "My rig is a thousand pounds overweight. I'm hauling two thousand pounds of pigeons—and so I've got to keep half of them in the air at all times." Okay, so it's a joke, but would that really work?

A very old joke indeed, but with an intriguing scientific hook.

No, it wouldn't work.

Think of it this way. The trailer is a box full of stuff. The box weighs so many pounds. Can banging on it possibly change its weight, whether it happens to be filled with gold bricks, sand, goose feathers, pigeons, or butterflies? Obviously not. The weight of a collection of material is the sum-total of the weights of the molecules in it, no matter how you rearrange them.

But what throws many people is the fact that airborne butterflies and pigeons are not resting on the floor, like other kinds of cargo. So how can their weight be transmitted to a scale that an inspector might place beneath the truck?

Through the air.

Air is, after all, a substance, albeit a thin and invisible one (see p. 151). It is made of molecules like everything else, and it therefore has weight: 1.16 ounces for each cubic foot at sea level, to be specific. The terrified pigeon who is catapulted into unplanned flight stays up in the air by repeatedly pushing down upon the air with its wings. (This is an oversimplification of bird flight, but it will do.)

When the wing presses down upon the air, the push is transmitted, molecule by molecule, throughout the air. (You would be able to feel the breeze if you were there, wouldn't you?) The pressed-upon air in turn presses upon everything it is in contact with, including the walls, floor, and ceiling of the trailer. The pigeon's wing-push force therefore remains completely within the trailer and doesn't change its effect upon a scale.

But, you might say, when the pigeon takes off, doesn't it push down on the floor of the trailer, making it instantaneously heavier, instead of lighter? And even after the pigeon is in the air, don't its downward wing thrusts constitute an extra downward force on the trailer via the air, again making it instantaneously heavier?

Right on both counts. But according to Sir Isaac Newton, every action has an equal and opposite reaction. Thus the downward push on the trailer is exactly cancelled by an equal upward push on the pigeon. Come to think of it, that's why it flaps its wings in the first place.

Perhaps what the truck driver should have done was to install a drain in the floor, introduce a cat into the trailer, and drain out the pigeon sweat as it accumulated.

NITPICKER'S CORNER:

No, pigeons don't sweat.

4

The Marketplace

From the street vendor to the glitziest mall, it's the same old jungle out there: people selling and people buying. The sellers always have the advantage because they know exactly what it is they're selling, while the *emptors* must constantly *caveat*. In many cases, the buyer not only doesn't know what the product really is, but can't even get a good look at it through the fog of promotion, packaging, and pitch.

In this section we will take a clear-eyed look at what some products really are, beneath the surface. We will visit the supermarket, the hardware store, the drugstore, and a restaurant, with a stop or two at the local pub.

A Real Cool Con

How do those "natural defrosting trays" work? They're supposed to take heat directly out of the air to defrost your frozen food quickly, yet they don't use any batteries or electricity.

Yes, and they're especially good at taking money directly out of your wallet. What they are is a remarkable new high-tech, space-age miracle called a *slab of metal*.

Of all types of materials, metals are the best conductors of heat. So if you put your frozen hamburger on a slab of metal, the metal will dutifully conduct heat from the warm room into the cold hamburger, and it thaws in a relatively short time. That's all there is to it. It's no more remarkable than the fact that any piece of metal feels cold to you ("cold steel") because it conducts heat from your warm skin into the relatively cooler room. Leaving frozen food out in the air is the slowest way to thaw it, because air is just about the worst heat conductor around.

The "miraculous, all-natural" defrosting tray, made of "an advanced, superconductive alloy," is nothing but a slab of aluminum. Aluminum conducts heat more than half as well as silver, which is the best conductor of all (see p. 35). Aluminum sells for about forty cents a pound, but you'll pay fifteen to twenty bucks for two pounds of it in the form of that miraculous defrosting slab.

Oh, yes, one little detail. The instructions tell you to "condition" the slab by running hot water over it for a minute or so before using it each time, and again half-way through the thawing process. Sounds like cheating to me.

But what hooks many people is the astounding demonstration that the slab manufacturers challenge you to try: Put an ice cube on the miracle slab and another one on the kitchen counter beside it. Lo and behold! The one on the slab melts quickly, while the one on the counter just sits there looking embarrassed. It really works.

What's going on? Well, the people selling the slabs are pretty sure that your kitchen counter is made of plastic laminate, tile, or wood—materials that are such poor conductors of heat that they're actually heat insulators. Naturally, the ice won't melt nearly as fast on an insulator as it will on a metal heat conductor. But try the demonstration this way: Put one ice cube on the slab and the other one on an unheated, heavy aluminum frying pan beside it. You'll find that they melt in exactly the same amount of time.

TRY IT *You can thaw your frozen food quickly by unwrapping it and placing it on an unheated, heavy* frying pan—or for even faster service, on one that you've warmed with hot water (not on the stove). Except for the cast-iron ones, frying pans are deliberately made to be good conductors of heat, so any heavy pan other than iron will work as well as those "miracle" trays. (Iron conducts heat only about one-third as well as aluminum.)

Of course, if you only had a big slab of solid silver . . . What about that sterling silver tea tray that you inherited from Grandma? It's 92.5 percent silver, and it will work twice as well as that overpriced chunk of aluminum.

A Foggy Day at the Bar

I must have opened thousands of bottles of beer. (No remarks, please; I'm a bartender.) Many times, as soon as I pop the cap, wisps of fog appear in the neck of the bottle, and sometimes even puff up above the opening. I've seen my share of foggy customers, but what causes foggy beer?

The fog is exactly the same as any fog: a collection of tiny particles of liquid water that have been condensed out of the air by a cold temperature, but are too tiny to fall down like rain. They are kept suspended by being constantly bombarded by air molecules. They look white because they reflect all wavelengths of light equally (see p. 42).

Your puzzlement apparently stems from the fact that you can't see any fog inside the bottle until you open it, yet it is equally cold at both times. What is there about opening the bottle that makes the fog form?

The space above the beer in the unopened bottle is filled with a mixture of compressed carbon dioxide, air, and water vapor—all gases. The water molecules in the vapor are content to stay that way—far apart from one another as an invisible gas, rather than clumping together as particles of fog.

They do so because they got there in the first place simply by leaping out of the beer's surface, and at the temperature of the beer, only a certain number of them, and no more, will have had enough energy to leap into the void (see p. 45). (Techspeak: The water vapor is in *equilibrium* with the liquid at that temperature.) The water molecules remain that way until you come along and upset everything by removing the cap and releasing the pressure.

When the pressure is released, the compressed gases are suddenly able to expand, and when gases expand, they lose some of their energy and are cooled (see p. 135). The gases are now cold enough to condense out some of the water, and that's the fog that you see.

If the bottle is put down in front of the customer and not poured immediately, you may see some of the fog actually rising above the mouth of the bottle and spilling over onto the bar. Dissolved carbon dioxide gas is now leaving the beer and expanding as it hits the warmer air at the top of the bottle. As it expands, it lifts some of the fog. Then, since carbon dioxide is heavier than air, it actually spills over like an invisible waterfall, carrying some of the fog down with it along the sides of the bottle.

No offense, but if you worked in a higher-class establishment you'd notice exactly the same fog effect upon opening bottles of champagne, and for exactly the same reasons.

Calories, Calories, Calories

The label of every food on the supermarket shelves tells us how many Calories there are in it. I know what a Calorie is—it's an amount of energy—but how do they determine how much energy a food will actually give you? Do they feed it to rats and then put them on treadmills to see how far they can run?

Let's not think of food energy as energy to be used only for exercising and running around. Our bodies utilize the energy we get from food not only to move, but to digest and metabolize the food itself, to repair the continuous day-to-day wear and tear on our cells, to build new growth, and to fuel the thousands of incredibly complex chemical reactions that keep everything balanced and working right. As evidenced by the multibillion-dollar weight-loss and diet industry, different individuals make use of their food's Calories in vastly different ways.

A Calorie, as the term is used by nutritionists, is the amount of energy it takes to raise the temperature of one thousand grams (a kilogram) of water by one degree Celsius. The chemist's *calorie* (small *c*) is one thousandth of the nutritionist's *Calorie* (capital *C*), but you generally won't see the capital *C* being used, except in this book. (See box on p. 65.)

People say that exercise "burns Calories." That's a very loose statement, of course. Energy doesn't burn; you can't set fire to it. But as any novice cook quickly learns, you *can* burn *food*. The energy in a food is released when the food burns, just as the energy in coal is released when we burn it. And that's how they determine the Calorie content of a food: They actually burn it and measure how many Calories of heat are released.

When we burn coal, the coal plus oxygen produces energy and carbon dioxide. Similarly, our bodies burn food—we call it metabolism—although much more slowly, and mercifully without the flames. (Heartburn doesn't count.) But the overall results are the same: Food plus oxygen produces energy and carbon dioxide. Remarkably, the amount of energy we get out of our food by metabolism is exactly the same as if we had burned it in a fire.

Nutritional technicians put a known amount of the dried food into a steel chamber full of high-pressure oxygen, immerse the whole thing in water, ignite the contents electrically, and measure the rise in the water's temperature.

From that they can calculate the number of Calories that were released: For each kilogram of water, each degree (Celsius) of temperature rise means that one Calorie of heat was released.

After setting fire to every food in sight, people eventually began to realize that every gram of protein gave off just about the same number of Calories, pretty much regardless of which protein it was or what particular food it was a part of. And the same went for fats and carbohydrates. They found that proteins and carbohydrates contain four potential Calories of energy in every gram, while fats contain nine Calories per gram. So these days, nobody bothers to burn the foods. Chemists analyze them for the number of grams of protein, fat, and carbohydrate, and then they calculate the total number of Calories from that.

Of course, everybody still burns marshmallows.

NITPICKER'S CORNER:

It's really rather surprising that when food and oxygen are converted into energy and carbon dioxide, it doesn't matter how that conversion is accomplished, whether by slow metabolism in a human being or by a blazing explosion inside a steel container in a laboratory. The energy released—the number of Calories—is the same, either way.

It's a general principle of chemistry: In any chemical process, if you start with chemicals in condition A and wind up with chemicals in condition B, the overall change in chemical energy will be the same, no matter how you got from A to B. We can liken amounts of energy to amounts of altitude: higher energy, higher altitude. If you start out on hill of altitude A and hike to a hill of altitude B, you've changed your altitude (your content of potential energy) by the amount B minus A, regardless of how roundabout your route may have been from A to B.

Corny, but Sweet

When I read the ingredients on the labels of prepared foods, I keep seeing "corn syrup," "high fructose corn syrup," and "corn sweeteners." But when I buy "sweet corn" in the markets, it's never very sweet, regardless of the seller's assurances. So how do they get all that sweetness out of corn?

You'll be the first to admit, won't you, that corn contains a lot of starch? Well, starch is the key to corn syrup. They convert cornstarch into sugar by the magic of chemistry.

Take away the water from a kernel of corn and the remainder is about 82 percent carbohydrates, a classification of natural organic compounds that includes sugars, starches, and cellulose. The cellulose, a tough material that makes up the cell walls of most plants, is in the corn kernel's skin. The sugars, as you already know, aren't very abundant. That leaves starches as the major constituent of corn kernels.

Bushel for bushel, the United States produces roughly five thousand times as much corn as sugar cane. And much of the sugar that we import comes from tropical countries that have never won awards for political stability or friendliness to the United States. So if American food producers could just make sugar out of cornstarch, they would be in great shape. Well, they can.

Sugars and starches are very close chemical cousins. In fact, starch molecules consist of hundreds or thousands of glucose molecules, all stuck together, and glucose is a fundamental sugar. So, in principle, if you could break corn's starch molecules down into smaller pieces, you would get a lot of loose molecules of glucose. You will also produce some molecules of maltose, another sugar whose molecules consist of two glucose molecules, still stuck together. And you would get a number of even larger fragments consisting of dozens of glucose units stuck together. Because these larger molecules can't flow past one another as easily as small molecules, the mixture you would wind up with would be thick and syrupy.

It turns out that almost any acid, as well as a variety of enzymes from plants and animals, can perform this trick of breaking starch molecules down into a syrup of various sugars. Enzymes in saliva do it all the time. (An enzyme is a natural substance that helps a specific chemical reaction to take place. Many important life processes wouldn't work without enzymes.)

TRY IT *Chew a starchy saltine cracker for several minutes and it will start tasting sweet.*

Glucose and maltose, however, are only about 70 percent and 30 percent as sweet, respectively, as sucrose, that wonderfully sweet sugar in sugarcane that we are in the habit of referring to as just plain sugar. So if you break down cornstarch the way we've described, it might average only about 60 percent of the sweetness of "real sugar." Food processors

get around this by using yet another enzyme to convert some of the glucose into fructose, a sugar that's even sweeter than sucrose. That's why you'll see "high-fructose corn syrup" listed on some labels.

But there's another problem. Glucose-maltose-fructose corn syrup may be a great economic boon to the American food industry, but it just doesn't taste quite the same or carry other flavors as well as good old sucrose. Fruit preserves and soft drinks, for example, just aren't what they used to be before the food processors abandoned cane sugar for the cheaper and more readily available corn sweeteners. As a label-reading consumer, the best you can do is to choose those products that are sweetened with the highest proportion of sucrose, which is listed simply as "sugar" on the labels.

By the way, if you can ever find a bottle of pre-1980 Coca-Cola, you will see what I mean. That's the year that Coke reportedly switched from sugar to corn sweeteners in its U.S. bottling plants. It is undoubtedly still made with sugar in countries where sugar cane is cheap. Next time you're south of the border, bring some back. But don't use the word Coke within earshot of a customs agent.

Danger! Exploding Matzos

Why do matzos have that corrugated-cardboard shape? Is it just traditional?

No, it's very practical.

Tradition, namely the dietary laws for the Jewish holiday Passover, demands the absence of any leavening agents such as yeast, baking soda, or baking powder. So matzos are made of flour and water only. Between Passovers, the manufacturers can cheat a little and add other ingredients to vary the flavor.

If you make dough by simply dumping flour and water

into a big mixer, you're bound to get some air bubbles beaten into it. Then when you roll the dough out thin and put it into a very hot oven—matzo bakeries use 800–900 degrees Fahrenheit or about 400–500 degrees Celsius—the trapped air will blow up and you will have an oven full of kosher shrapnel.

So before baking the rolled-out dough, the matzo bakers run it through a stippler, a gang of spiked wheels that roll over the dough and puncture the air bubbles. It's the stippler that puts those perforated furrows in the matzo. You'll still see some small blisters between the stippler tracks, but they're less-than-destructive in size and they contribute to an interesting appearance by browning faster than their surroundings.

TRY IT Examine a cracker—of any religious persuasion—and you'll see that it has been punctured with a pattern of holes.

The reason is the same as for matzos: to prevent blistering in the oven. The puncturing of ordinary crackers doesn't have to be so radical, however, because the dough is leavened, which makes very tiny gas bubbles (see p. 60), and it isn't baked at such an extreme temperature.

Blowing Hot and Cold

When I sprained my ankle playing softball, somebody ran to a drugstore and bought a cold pack. They squeezed it and shook it, whereupon it turned into an instant cold compress. What's inside that package that makes it get cold so fast?

The cold pack contains ammonium nitrate crystals and a thin, breakable pouch of water. When the pack is squeezed, the water pouch breaks and, with a little shaking, the ammonium nitrate dissolves in the water.

When any chemical dissolves in water, it may either absorb heat—get cold—or release heat—get hot. Ammonium nitrate is one of those that absorb heat. It takes the heat right out of the water, thereby cooling it. And the amount of cooling isn't trivial. That cold pack can actually get down close to freezing.

Because doctors keep blowing hot and cold about when to apply heat to an injury and when to apply cold, there are almost as many hot packs on the market as there are cold packs. The hot packs contain one of those chemicals that give off heat when they dissolve in water, usually crystals of calcium chloride or magnesium sulfate.

But why should a chemical absorb or release heat during the simple process of dissolving in water? After all, at home we dissolve crystals of two common chemicals, salt and sugar, in water time after time, yet we never see the sugar, for example, cooling off our hot coffee or heating up our iced tea. The fact is that salt and sugar are exceptions (see below).

When a chemical substance dissolves in water, it is a two-step process: first, the chemical's solid, crystalline structure is broken down, and then a reaction takes place between the water and the broken-down chemical parts. The first step invariably has a cooling effect, while the second step has a heating effect.

If step one cools more than step two heats, as in the case of ammonium nitrate, the overall effect is cooling. If it's the other way around, as it is with calcium chloride and magnesium sulfate, the overall effect is heating. In the cases of salt and sugar, the two steps happen to be just about equal, so they cancel each other out and there is very little change in temperature.

NITPICKER'S CORNER:

Here's what's going on during the two-step process of solid crystals dissolving in water:

A crystal is a rigid, three-dimensional, geometric arrangement of particles. The particles may be atoms, ions (charged atoms), or molecules, depending on the substance we're talking about. We'll just call them particles.

Step 1: The particles must first be released from their rigid positions in the crystal in order to be able to float about freely in the water. To break down any rigid structure requires the expenditure of energy; somebody or something has to supply the sledgehammer blows that knock the structure apart. During the breakdown of the crystal's structure, therefore, some heat energy must be borrowed from the water, and the water cools down accordingly.

Step 2: The liberated particles don't just swim around in splendid isolation. They have a strong mutual attraction for water molecules (see p. 48). If they didn't, they wouldn't have been interested in dissolving in the first place. So as soon as they are in the drink, they are literally attacked by water molecules, which rush to cluster around them like floating magnets around a submarine. When magnets (or molecules) are attracted to something, they expend energy in the rush toward their targets. This energy heats up the water.

Now it's just a matter of which effect is bigger: the cooling effect from the breakdown of the solid or the warming effect from the particles' attraction for water molecules. If the cooling is bigger, the net effect will be that the water gets colder when the solid dissolves. That's how it is with ammonium nitrate. If the warming is bigger, the net effect is that the water gets warmer when the solid dissolves; that's how it is with calcium chloride and magnesium sulfate.

Salt and sugar? It's just an accident that the two effects are approximately equal and cancel each other out. So there is practically no net cooling or heating when salt or sugar dissolves in water. (Actually, salt—sodium chloride—does cool the water very slightly when it dissolves.)

TRY IT *Ammonium nitrate is a common fertilizer and calcium chloride is a common drying agent, sold for drying out damp closets and basements. You may have these chemicals around the house or farm. Stir ammonium nitrate into water and the water will get very cold. Stir calcium chloride into water and it will get quite hot. (Don't cover and shake; the heat can make the liquid splatter.) A couple of tablespoons of the solid in a glass of water will do.*

Freeze, Baby, Burn!

Who ever came up with that ridiculous oxymoron, "freezer burn"? And what has actually happened to "freezer-burned" food?

Sure, it's an oxymoron, but not a bad one at that. Take a good look at that ancient pork chop you've been keeping in the freezer for an emergency. Doesn't its parched and shriveled surface look as if it had been seared? Believe it or not, "seared" doesn't only refer to heat; it also means withered or dried out. And that's exactly what the freezer has done to the meat: dried it out.

How does coldness make food dry out and wither? Please observe that the patches of "burn" on that forlorn pork chop are dry and parched looking, as if much of the water had been sucked out. And it has. But in what state or condition does the water exist in frozen foods? Correct. It is indeed ice. So we are forced to the conclusion that while the hapless chop in question has been languishing in your freezer for much longer than you had ever believed possible, something has been spiriting away ice molecules (which are water molecules, of course) from its surface.

But how can water molecules, so firmly anchored in solid ice crystals, be wafted off to some other location? It turns out that they will spontaneously migrate, if they can, to any place that offers them a more hospitable climate. To a water molecule, that means a place that's as cold as possible, because that's where its energy is lowest, and, other things being equal, Mother Nature will always try to find the lowest-energy conditions for her little particles. (You can drive water molecules away by heating them, right?)

So if the wrapping on the food isn't absolutely water-molecule-tight, water molecules will migrate through it or around it, from the ice molecules in the food's surface to any other location that's colder, such as the walls of the freezer itself. The net result is that water will leave the food and be disposed of by the freezer's defrosting mechanism. The food's surface is left parched, wrinkled, and discolored.

This doesn't happen over night, of course. It's a slow process, and it can be slowed down to practically nothing simply by wrapping the food tightly in a material that is impermeable to wandering water molecules. Some plastic wraps do a better job than others. Best of all are vacuum-sealed, thick plastic packages like Cryovac, because in addition to being quite impermeable to water vapor, they leave no space between the food and the wrapping. If there is any air space in a package, water molecules will migrate across it to the inner wall of the wrapping and settle there as ice, for the same reason that they'd go all the way to the freezer walls if they could. And you'll still get freezer burn.

Moral Number 1: For long-range keeping of frozen foods, minimize freezer burn by: (a) using a wrapping material specifically designed for freezing because of its impermeability to migrant water molecules and (b) wrapping the food very tightly, leaving no air pockets.

Moral Number 2: When buying already-frozen foods, watch for any ice crystals in the space inside the package. Where do you think that water came from? Right. The food. So either it's been "burned" by being kept too long or it's been thawed, which releases juices from the food, and refrozen. Shop somewhere else.

Oysters on the Half-Shelf

Half the calcium supplements on the health-food shelves seem to be ground-up "natural oyster shell." Is oyster-shell calcium better than other kinds?

If Gertrude Stein had been a chemist, she might have said, "Calcium carbonate is calcium carbonate is calcium carbonate."

Sure, clams and oysters make their shells out of calcium carbonate. But chemically speaking, it doesn't matter whether the calcium carbonate in the supplement bottle came from an oyster bed or a bed of limestone, which is also made of calcium carbonate. Neither is more "natural" (whatever that means) than the other. Oysters incorporate a bit of nonmineral matter in their shells, however, so calcium carbonate from other sources might be a bit purer.

Calcium supplements are sold in other chemical forms besides calcium carbonate (read the labels). But weight for weight, these other forms contain less calcium than calcium carbonate does, and it's the actual element *calcium* that you're after; never mind the rest of the stuff. Calcium carbonate contains 40 percent calcium by weight, while calcium citrate contains 21 percent, calcium lactate contains 13 percent, and calcium gluconate contains only 9 percent calcium.

Now you can figure out which supplement on the shelf gives you the most actual calcium for your money.

Savory, but Not Flavory

What, exactly, is MSG, and what does it do to foods? It's billed as a "flavor enhancer," but how can just adding a substance to a food improve its flavor no matter what that flavor is?

It does sound odd, but something is really going on here.

What makes the MSG story hard to swallow is that the

terminology is misleading: "Flavor enhancers" do not enhance the flavors of food in the sense of improving them; that is, they don't make anything taste *better*. What they do is intensify, or magnify, the flavors that are already present—regardless of whether those flavors are delicious, indifferent, or downright repulsive. The food processing industry prefers to call them "potentiators." We'll call them *flavor intensifiers*.

How do they work? Some flavor experts talk in terms of *synergism*, a situation in which the total effect of two things acting together is greater than the sum of their effects when acting individually. In other words, the whole is greater than the sum of its parts. A flavor intensifier may have little or no flavor of its own, but when it is combined with something that does have a flavor, that flavor is perceived as being stronger than it would have been by itself.

Exactly how the intensifier tricks our taste buds into giving us a more intense sensation is still being tracked down by researchers. One theory is that the intensifiers help certain flavor molecules to stick to the receptor sites on our tongues longer or more tightly. MSG seems to have a particular talent for intensifying salty and bitter flavors.

MSG is monosodium glutamate, a derivative of glutamic acid, which is one of the common amino acids that proteins are made of. It isn't the only flavor intensifier, however. Two other chemicals that act in this way are known in the trade as 5′–IMP and 5′–GMP (chemists call them disodium 5′–inosinate and disodium 5′–guanylate). All three are derivatives of natural amino acids that occur in vegetation such as mushrooms and seaweeds.

The flavor-brightening qualities of these plant materials have been known for thousands of years. The Japanese, for example, traditionally use seaweed in subtle, delicate soups that can benefit substantially from a flavor boost. Japan is the world's major producer of pure MSG, a white crystalline powder that has been sold by the ton for decades. Its major use is in the manufacture of prepared foods, although Chinese restaurants often use it as an off-the-shelf ingredient in cooking.

Recently, MSG has been suffering a bit of a bad rap because some people have unpleasant reactions to it. All evidence seems to indicate that the problem, if it can be called that, is that certain people are ultrasensitive to MSG, not that there is anything inherently harmful about MSG itself except when taken in huge doses. But almost anything is harmful in huge doses.

The FDA (U.S. Food and Drug Administration) has not yet required the separate listing of MSG content on food labels. But you may find it, or its close chemical cousins, on the labels of soups and snack foods hiding behind a variety of aliases, including Kombu extract, Glutavene, Aji-no-moto (on Japanese products), and hydrolyzed vegetable protein, which is plant protein that has been broken down into its constituent amino acids, including glutamic acid.

A wide variety of other flavor-intensifying compounds are extracted from yeasts. One company makes and sells to food processors over two dozen yeast-based "flavor enhancers," specifically tailored to intensify certain flavors, from beefy to chickeny to cheesy and salty. You'll see them listed among the ingredients on the package as "yeast extract," "yeast nutrient," or "natural flavor," even though, strictly speaking, they're not flavors. On the other hand, they're not MSG either.

Not Bloody Likely

I like my steaks real rare. But what do I say to those holier-than-thou types who taunt me about eating "bloody" meat?

Nothing. Just smile. They're wrong.

They're not necessarily wrong in their preference for eating well-done steaks, although many would argue that such behavior should be declared a felony. Where they're wrong is in calling your steak bloody. There's practically no blood in it at all.

Blood, you might politely remind them, is that red liquid that circulates through the veins and arteries of a living animal. Not to get too gross about it, but in the slaughterhouse, just as soon as the animal is zonked, they drain out virtually all of the blood except for what remains trapped in the heart and lungs, which, we devoutly hope, are of minimal gastronomical interest to you and your friends.

When you order a steak, you're ordering muscle tissue, not circulatory-system components. Blood is red because it contains *hemoglobin*, an oxygen-carrying protein. The red color of muscles, however, is due to a chemical compound called *myoglobin*, a protein that stores up oxygen on the spot, right in the muscles, for whenever they might need it for a sudden burst of energy. It is only a coincidence that both of these compounds are red and that they both turn brown when cooked. (Well, there are really no coincidences in nature; everything happens for a reason. The reason in this case is that hemoglobin and myoglobin are very similar iron-containing proteins.)

Various meats contain various amounts of myoglobin, because various animals have varying degrees of need to store oxygen in their muscles for bursts of energy. Pork (those lazy pigs!) contains less myoglobin than beef, chicken has even less, and fish has less still (see p. 56). So there are red meats and there are relatively white meats. Ask your friends to explain *that* in terms of bloodiness.

BAR BET A "bloody-rare" steak has no blood in it.

Shake It, But Don't Break It

This is a question in classical mechanics. What is the best way to get ketchup out of the bottle?

The very best way, as was once memorably demonstrated by David Letterman, is to grasp the bottle firmly around the bottom and swing it round and round over your head, like a lariat. Of course, the ketchup will splatter all over the walls, but you asked only how to get it out of the bottle, didn't you?

There is one method that doesn't stand a prayer of a chance, yet you see people trying it in restaurants all the time: pounding on the bottom of the bottle. All that does is make Isaac Newton rotate in his crypt at Westminster Abbey. Sir Isaac taught us (or at least he thought he did) his three laws of motion, the basic laws of mechanics that govern how things move. Had he known about ketchup (which arrived in England only around the time of his death in 1727), he would have stated a fourth law, thus: "He who slammeth ye bottom of ye ketchup bottle doeth naught but drive bottle even more tightly onto ye ketchup." Or, since according to Sir Isaac every action hath an equal and opposite reaction, you're just driving the ketchup more tightly into the bottle—just the opposite of what you want to accomplish.

On the other hand, the Letterman method is legitimately Newtonian because you are applying an outward centrifugal force to both bottle and ketchup, yet restraining the bottle (we hope) while allowing the ketchup to react freely to the outward force.

So how would Sir Isaac advise you to de-bottle the sauce without attracting quite so much attention? There are two ways.

First, you can slightly modify the centrifugal force method by holding the bottle horizontally and snapping your wrist downward, so that the tip of the bottle describes a short downward arc—a portion of a circle. As in the Letterman method, the ketchup will experience a centrifugal force outward from the center of the circle and hence outward through the neck of the bottle, a direction which we devoutly hope is toward your plate rather than toward a dinner companion. The latter unfortunate circumstance can result if you begin the arc a trifle too high above the plate.

TRY IT *The second, and safer, Isaac-approved ketchup-removal technique is to give the upside-down bottle a brisk, downward, stabbing-like thrust along its axis, aiming directly at the plate but stopping short at the last possible instant.*

In this maneuver, the ketchup inside the bottle will be tricked. It will keep moving toward the plate even though the bottle itself has stopped, just as a driver will move toward the windshield when his car is stopped short by a telephone pole. Or, to paraphrase Sir Isaac, "A body in motion will continue in motion until stopped by a windshield or a french-fried potato."

If the bottle is new or has been recently refilled (restaurants do that), first loosen the ketchup by rotating a knife blade inside the neck.

The only remaining problem is that, seated at a table, you may not have enough thrusting distance above the plate to carry out a good, swift bottle-stab. So stand up.

That's Oil, Folks

The ingredient lists on margarine and other food packages say "partially hydrogenated vegetable oil." If an oil has to be hydrogenated, whatever that means, why is it only partially?

Hydrogenation is, quite sensibly, the name of the process by which hydrogen is added to something. Hydrogen is the lightest of all known substances, but paradoxically, hydrogenating an oil makes it thicker and more solid. If we went all out with complete hydrogenation, the oil would become as solid as candle wax, which would make your margarine rather hard to spread.

Oils, whether from plants or petroleum, are made of molecules that can have "bonding gaps" between the atoms—not actual gaps in space, but regions of incomplete chemical-combining power (Techspeak: double bonds). At these locations in the molecules the yearnings of atoms to join with other atoms are not fully satisfied. The atoms still have some unused bonding power that they could use to grab other atoms, if only the right other atoms would come along. (Techspeak: Such unfulfilled molecules are said to be *unsaturated*. If there is only one unfulfilled location in the molecule, it is called *monounsaturated*.)

Hydrogen is the perfect candidate for consummating the bonding desires of these unfulfilled atoms. It is the smallest atom of all, and can snuggle into almost any place in a convoluted molecule where it may be needed, especially if it is forced in under high pressure, which is how they hydrogenate oils. By filling in the gaps, the hydrogen atoms thoroughly satisfy the molecules' longing to form bonds. (Techspeak: Fully bonded molecules are said to be *saturated*.)

What does that do to the oil? The saturated molecules are more compact once their bonding gaps are filled in, because they are somewhat more flexible. (Techspeak: Double bonds are more rigid.) They can therefore nestle together more tightly into their solid form, and the substance will stay solid longer when heated. That is, it will melt at a higher temperature. (Whenever an oil happens to be a solid at room temperature, we call it a fat instead of an oil. In fact, the technical term for them all, whether liquid or solid, is *fat*.)

We want to make vegetable oils semisolid for use as spreadable margarines, for example, but we don't want to make the margarines too hard. That was no joke about the candle wax. The paraffin in a candle is actually a mixture of completely saturated oils; they come from petroleum, however, rather than from plant seeds.

In general, vegetable oils tend to be mostly unsaturated and are liquid at room temperature, while animal fats tend to be mostly saturated and solid. Vegetable oils contain about 15 percent saturated molecules. To make margarine, they are partially hydrogenated up to about 20 percent. Butter is about 65 percent saturated.

Unsaturated oils tend to break down and smoke at relatively low sautéing temperatures. They also turn rancid rather easily, because oxygen molecules from the air can get into those gaps between the atoms and attack them. Hydrogenation makes the oils more stable because it plugs the gaps with hydrogen atoms.

That's the good news about hydrogenation. The bad news is that saturated fats appear to raise people's blood cholesterol and increase their risk of heart disease. Food manufacturers are engaged in a never-ending struggle to keep saturated fats to a minimum so that they can brag about their products' healthfulness, while at the same time hydrogenating their oils enough to give them desirable properties.

The Great Fog Forgery

Why is dry ice dry? And what makes all those clouds of smoke around it?

It's not smoke; it's fog. Although dry ice is pure carbon dioxide, the fog itself is not carbon dioxide, as some people think. Carbon dioxide gas is invisible. The cloud of fog surrounding the dry ice is pure water. It has been condensed out of the air's natural humidity by the dry ice's low temperature.

Dry ice is carbon dioxide in the solid form, just as regular ice is water in the solid form. Water freezes to a solid at 32 degrees Fahrenheit (0 degrees Celsius), while carbon dioxide doesn't become solid until 109 degrees below zero Fahrenheit (–78.5 degrees Celsius). Thus, frozen carbon dioxide (dry ice) is much colder than frozen water.

Regular ice is wet because as it melts it becomes liquid water. Dry ice is dry because it doesn't melt. It changes directly into a gas without turning into a liquid first. Carbon dioxide simply cannot exist as a liquid at ordinary atmospheric pressure. When it finds itself in the even more unnatural solid form as dry ice, it does its best to revert directly to a gas.

Carbon dioxide is most comfortable as a gas because in a gas the molecules are as far apart as they can get, and carbon dioxide's molecules do not like each other very much. That is, they don't stick together very well the way water molecules do (see p. 78). In liquids, the molecules are always somewhat stuck together as they slither around, over, and under one another. But carbon dioxide molecules just don't have the necessary stickiness to be a liquid, unless you force them so close together that they have no choice but to associate. In other words, carbon dioxide gas will become a liquid only under duress: under high pressure. Carbon dioxide is shipped around the country this way—as a liquid under high pressure in steel tanks.

CO_2 fire extinguishers are nothing but tanks of liquid carbon dioxide with a squeeze valve. When you release some of the pressure by squeezing the trigger, carbon dioxide comes rushing out through the funnel. It comes out as a blast of very cold gas (see below), mixed with a "snow" of solid carbon dioxide. If you could collect enough of this snow before it evaporates and squeeze it into a "snowball," you'd have a chunk of dry ice. On a larger scale, that's exactly how they make dry ice from liquid carbon dioxide.

The fire extinguisher works in two ways: The coldness can lower the temperature below the fuel's ignition point, while the carbon dioxide smothers the fire because it's a heavy gas that pushes away the oxygen (see p. 213).

Dry ice is used on movie sets to make fog. It is real fog because it consists of microscopic droplets of water suspended in the air. But you can always tell a fog forgery, because the water is very cold from the dry ice and the fog therefore lies on the ground like a blanket—unless it is blown around by an off-camera fan. Real, weather-generated fog, on the other hand, hangs fairly motionlessly in the air.

Movies also use dry ice to simulate caldrons of boiling water. Just throw some dry ice into the water, and as the solid carbon dioxide changes to gaseous carbon dioxide, the gas rises through the water as fog-filled bubbles that break at the surface and are supposed to look like steam. If you look closely, though, you can always tell that it's fake. The microscopic droplets in fog reflect light and look white, but steam is made of bigger droplets of water that are almost transparent. Moreover, steam goes straight up because heat rises, while the cold dry-ice fog hangs low over the caldron.

While we're on the subject of movie fakes, how about those scenes of storm-tossed ships? Are they just miniature models, shot at slow motion in a big tank? There's a sure-fire way to tell. Check the size of the water droplets from the crashing waves. If they're the size of a porthole or a cannon ball on the ship, it's a model in a tank. Water just doesn't break up into drops the size of cannon balls, unless the "cannon balls" on the ship are really BBs on a scale model.

You didn't ask, but . . .

Why is the blast of snowy gas that comes out of a CO_2 fire extinguisher so cold, even though the extinguisher may have been sitting around in the room for months?

When the liquid carbon dioxide in the tank turns into gaseous carbon dioxide, it gets cold enough to freeze some of the carbon dioxide into "snow." But why, indeed? All that's happening is that the compressed carbon dioxide gas is expanding as we let it loose into the room. Do expanding gases automatically get cold? Yes, they do, and here's why.

The molecules in an expanding whoosh of gas have the power to blow things away, don't they? Notice the powerful blast that you get when you use that fire extinguisher. If you're not careful, you can blow the whole, still-burning fire into the next county. So the molecules of an expanding gas can knock objects away—even if the object is only air—by banging up against the object's molecules. (What else is there to bang against?)

As the gas molecules knock an object's molecules for a loop, they expend some of their own energy and slow down, just as a billiard ball moves more slowly after it has hit another ball. And slower gas molecules mean lower-temperature gas molecules (see p. 236).

The carbon dioxide in the extinguisher tank is at such a high pressure that when you release it into the room it expands tremendously, with a correspondingly tremendous decrease in temperature.

The Old Skin-and-Bones Game

Somebody tried to tell me that clear, shimmering, sparkling-bright Jell-O, the tranquilizing treat of my childhood, is

made from pig skins, cow hides, bones, and hoofs. Yuck! Can that possibly be true?

Of course not. Just the skins and bones. No hoofs.

Jell–O and similar desserts are about 87 percent sugar and 9 or 10 percent gelatin, plus flavoring and coloring. Kids love three things about the stuff: It is brightly colored, it is very sweet, and it jiggles. Mothers don't mind, because gelatin is pure protein.

The gelatin, which of course is the jiggler, really does come from pig skins, cattle hides, and cattle bones. But stop squirming. Every time you've made soup or stock that jelled in the refrigerator, you've made gelatin out of chicken hides or beef bones.

The skin, bones, and connective tissue of vertebrate animals contain a fibrous protein called *collagen*. There is no collagen in hoofs, hair, or horns. When treated with hot acid (usually hydrochloric or sulfuric acid) or alkali (usually lime), collagen turns into gelatin, a somewhat different protein that dissolves in water. The gelatin is then extracted into hot water, boiled down, and purified.

You wouldn't want to see—or smell—the early stages of the purification process. But by the time the gelatin leaves the factory, it has been thoroughly washed at various stages to get the acid or alkali out, then finally filtered, deionized (a way of removing chemical impurities), and sterilized. What eventually leaves the factory is a pale yellow, brittle, plastic-like solid in the form of ribbons, noodles, sheets, or powder. When this solid gelatin is soaked in cold water, it absorbs water and swells up; then, when the water is heated, it dissolves to form a thick liquid that jells on cooling.

As a protein, gelatin is obviously nutritious, although it isn't what nutritionists call a complete protein. But what is most fun about it is that when dissolved in water it's a gel (a jelly-like substance) when cold and a liquid when warm. It literally "melts in your mouth." That's the main characteristic that it gives to confections from marshmallows to Gummi Bears, whose gumminess comes from a high proportion, around 8 or 9 percent, of gelatin. And guess what holds those tiny white dots on the tops of nonpareil chocolate candies? Right, gelatin used as an adhesive.

Most of the gelatin made in the United States—more than a hundred million pounds a year—is slurped in the form of gelatin desserts. You'll also find it in soups, shakes, fruit drinks, canned hams, dairy products, frozen foods, and bakery fillings and icings. But food isn't the only use for this unique substance. Those little two-piece capsules that contain many drugs are also made of gelatin—about 30 percent gelatin in 65 percent water. Match heads are a mixture of chemicals held together by a gelatin binder.

And then there is photography. The photographic emulsion—that thin, light-sensitive coating on the film or paper—is made of dried gelatin containing the light-sensitive chemicals. Nothing better than gelatin has ever been found since it was first used in photography in 1870. Isn't it heartwarming to know that astronauts take pictures using a primitive substance made of animal skins and bones?

There's Something Fishy Around Here

Why does fish smell fishy?

Silly-sounding question, maybe, but with several interesting answers.

People tend to put up with fishy-smelling fish in markets and restaurants because they're thinking, Well, what else should it smell like? But fish needn't smell like fish at all. Not if it's perfectly fresh.

When they're only a couple of hours removed from the water, fish and shellfish have virtually no odor. A fresh "scent of the sea," perhaps, but certainly nothing the least bit unpleasant. It's only when seafood starts to decompose that it takes on that "fishy" aroma. And fish decomposes much faster than other kinds of meat.

Fish flesh—fish muscle—is made of a different kind of protein from, say, beef or chicken (see p. 56). It breaks down more quickly, not only in cooking, but also under the action of enzymes and bacteria. In other words, it spoils faster. That fishy smell comes from the products of decomposition, notably ammonia, various sulfur compounds, and chemicals called amines that result from the breakdown of amino acids.

The human nose is remarkably sensitive to these chemicals. The odors are noticeable long before the food gets downright unhealthy to eat, so a slight fishy smell indicates only that the fish isn't as fresh—or as enjoyable—as it could be, not necessarily that it's dangerous.

Amines and ammonia are bases, which are counteracted by acids. That's why lemon wedges, which contain citric acid, are often served with fish. (If you buy scallops that smell a trifle ripe, rinse them in lemon juice or vinegar before cooking. But don't let them soak, because scallops absorb water like sponges and will then steam themselves when you try to grill or sauté them.) The very best way to test seafood for freshness is to ask as politely as possible to sniff the merchandise before buying, although in certain Mediterranean markets with scrupulously high standards, this can be taken as a grave insult.

A second reason why fish spoils more quickly than other meats is that, in the wild, most fish have the unfriendly habit of swallowing smaller fish. (It's a jungle down there.) They are therefore equipped with digestive enzymes that are exquisitely effective digesters of fish flesh. After the fish is caught, if any of these enzymes should escape from the guts through rough handling, they'll quickly go to work on the fish's own flesh. That's why gutted fish will keep longer than whole ones.

A third reason: The decomposition bacteria in and on fish

are more efficient than those on land because they're designed to operate in the cold seas. Warm them up slightly, and they'll really go to town. To stop them from doing their dirty work, we have to cool the fish down a lot more quickly and thoroughly than we do to preserve warm-blooded meat.

That's why ice is the fisherman's best friend—lots and lots of it. Ice not only lowers the temperature, but it keeps the fish from drying out. Fish don't appreciate being dry, even after they're deceased.

Reason number four: In general, fish flesh contains more unsaturated fats (see p. 131) than land-animal flesh. That's one of the reasons why we value it in these cholesterol-dreading times. But unsaturated fats turn rancid (oxidize) much more readily than those saturated fats that are so delicious in beef. The oxidation of fats turns them into foul-smelling organic acids, which contribute further to the unappealing aroma.

If you walk into a seafood restaurant and it smells fishy, depart immediately in search of the nearest hamburger.

The Proof Is in the Drinking

Wine and whiskey labels state the alcoholic strength as "proof" or as "percent alcohol by volume." Where does the term "proof" come from, and what does "by volume" mean?

The term *proof* was coined in the seventeenth century when people proved, or tested, the alcohol content of whiskies by moistening gunpowder with it and setting it afire. (Honest.) A slow, even burn indicated the desired 50-percent-or-so of alcohol. If the booze was watered down, the flame would sputter.

Today in the United States, 50 percent alcohol (by volume) is defined as 100 proof, so the proof is always twice the percentage of alcohol. An 86 proof gin, for example, contains 43 percent alcohol by volume. (In the United Kingdom the system is somewhat different; 100 proof is defined as 57.07 percent alcohol by volume, for reasons too painful to explain.)

Not having any gunpowder handy, how should we express the amount of alcohol in a liquor? The obvious way would be to quote a percentage: If the beverage is exactly half alcohol, we would say that it is 50 percent alcohol. But then some wise guy would ask, "Fifty percent of what? Do you mean 50 percent of the liquor's *weight* is alcohol, or 50 percent of its *volume* is alcohol?" And we wouldn't know what to answer because the weight percent and the volume percent (the percent by volume) can be quite different, especially where alcohol and water are concerned. There are two reasons for this.

First of all, alcohol is lighter than water. In technical language they have different *densities*. A pint of pure alcohol weighs only 79 percent as much as a pint of water.

Let's say that we wanted to make a 50 percent mixture of alcohol and water by weighing out equal weights—pounds or grams—of these two liquids and mixing them together. We would find that we have to use a bigger volume of alcohol than water. By *weights*, the mixture will certainly be 50 percent alcohol, but by volumes it will be *more* than 50 percent alcohol. (It works out to be about 56 percent.)

Now guess which type of percentage the beverage manufacturers have chosen to quote on their labels. Right. The one that makes the alcoholic content seem higher: percentage by volume. Taxes are usually based on the percentage of alcohol, so the tax man also profits by this dodge.

The second reason that percentage by volume is the chosen measure for wines and liquors is that a very unusual thing happens when alcohol and water are mixed: The final mixture takes up less space than the sum of the volumes that were mixed. In other words, the liquids shrink. Mix a pint of alcohol with a pint of water, and you get only 1.93 pints of mixture instead of the two pints that you would expect. The reason is that water molecules and alcohol molecules form hydrogen bonds (see p. 98) to each other, which snuggles them together even more tightly than they had been in the pure alcohol and water by themselves.

As you can imagine, this screws up the concept of percentage of alcohol by volume. Should it be the percentage of the volumes before mixing, or the percentage of the final volume after mixing? The beverage producers have decided to use the smaller volume, the volume *after* mixing. That's quite appropriate, of course, because that's the way we buy the product: already mixed. But if you are not too deeply mired among the mathematically challenged, you will quickly realize that this method of reckoning gives an even higher value for the percentage of alcohol. Using the beverage producers' method, that 50-percent-by-weight mixture that we so carefully weighed out would appear on the label as approximately 57 percent by volume.

BAR BET I can mix a pint of alcohol with a pint of water, and get a mixture that is more than 50 percent alcohol by volume.

You didn't ask, but . . .

When health experts talk about the benefits and dangers of alcohol, they talk in terms of how many actual grams of ethyl alcohol (grain alcohol) a person is consuming. How can I tell how many grams of alcohol there are in a drink?

Multiply the number of ounces of booze or wine in the drink by its percentage of alcohol by volume (one-half its proof). Then multiply that result by 0.233. The result will be the number of grams of ethyl alcohol in the drink. For example, a one-and-a-half-ounce shot of 80-proof whiskey contains 14 grams of alcohol ($1.5 \times 40 \times 0.233 = 14$).

5

The Great Outdoors

Step outside with me, please. Look beyond everything that has felt the hand of man. Look at the air, the sun in the sky, the clouds. And wonder at it all. How can something as insubstantial as air exert a "barometric pressure" on us? Why does the sun feel hotter at certain times of day? Why are some clouds black? Why does it get warmer when it snows? And if you've ever been to the beach, haven't you wondered why the waves always roll in in the same way, whether the coastline faces north, east, south, or west?

To paraphrase Charles Dudley Warner (and no, it wasn't Mark Twain), nobody does anything about the weather, but we sure can talk about it. Even better than talking about it, though, is *understanding* it, by observing the weather closely and thinking things through. In this section, we will experience sunshine, clouds, wind, and snow. And along the way we will comment on a couple of man-made outdoor phenomena: the Statue of Liberty and Fourth of July fireworks.

By the Beautiful Sea

Every time I'm at the seashore, there seems to be a cool breeze blowing in from the ocean. Is it my imagination, or is there something about the shore that makes it inherently cooler and windier?

You're right. "Sea Breeze" isn't just the name of a thousand beach motels. The breeze coming in from the sea is a real phenomenon that makes the shore cooler than it is inland—at least in the afternoon, which is when people most want to cool off anyway (see p. 145). In the daytime, cool breezes almost invariably blow in from the ocean toward the land, rather than the other way around. They begin several hours after sunrise, reach their peak by midafternoon, and die out toward evening.

Here's how it happens. Starting in the morning, the sun beats down on both land and sea. But the sea isn't noticeably warmed by the sunshine because it is so cold and vast that it has an inexhaustible appetite for heat energy, slurping it up with nary a degree's rise in temperature. The land, on the other hand, is substantially warmed up by the sun's rays. Soil, plant leaves, buildings, roads, and so on are relatively easy to heat. (Techspeak: they have low *heat capacities*, compared with water.) As the land warms up, it warms the air above it, which expands and rises. The cooler, denser air that is sitting over the water then rolls in underneath it, sweeping over the beach and cooling bathers.

It's not just that the ocean breeze is cool. Even if it weren't, it would still help to cool the sweltering hordes by evaporating perspiration. (See p. 177).

Doin' the Wave

At the seashore, why do the waves always break parallel to the shore, regardless of the direction in which the shoreline runs?

Waves can tell when they're approaching a shore and actually turn to line up with it.

What makes waves, of course, is wind blowing across the water's surface. But it can't be that the wind is always blowing the waves straight in to shore. Out in the ocean, the wind may be blowing every which way. The waves we observe at the shore are only those that are traveling more-or-less generally in our direction, or else we would never see them. Nevertheless, most of them approach obliquely, not straight in to the shoreline. What happens then, believe it or not, is that the incoming wave "feels" the shore and turns to face it squarely before breaking. Later, when it breaks (see below), the line of foam will be quite parallel to the shoreline.

The question, of course, is, how does a wave know when it is approaching a shore? And what makes it turn?

When a wave—consider it a broad bump on the surface—is still over deep water, there is nothing to restrain it; it goes wherever the wind commands. But as it moves into shallower water, the lower part of the wave begins to drag on the bottom, which slows it down. That's its clue that it is approaching a shore, and that "knowledge" gives it a preferred direction.

Let's say we're riding a wave that is coming in at an angle, with the shoreline at our left. The first part of the wave to hit shallow water and scrape bottom will be the left end of the wave. The left end will therefore be slowed down, while the middle and right ends keep going at the same speed. This has the effect of turning the wave to the left—toward the shore. (If you drag your left foot from a go-cart, you're going to swerve to the left, aren't you?) This dragging and slowing proceeds down the wave's length from left to right as more and more of the wave feels the drag, and gradually the whole wave is turned to the left. Its crest line is now stretched out parallel to the shoreline, and that's the position it finds itself in when it gets close enough onto the shore to break.

Waves break because of the same bottom-dragging effect. After the wave has lined up parallel to the shore, it eventually reaches such shallow water, and its bottom is slowed down so much, that its top overtakes its bottom and tumbles over it. The top falls with a crash, churning up a line of foam all along the wave's length—and that's parallel with the shoreline.

TRY IT The next time you fly over a curving coastline, notice how the white lines of foam from the breaking waves are always parallel to the shore, no matter which way the shoreline turns.

Ever on Sun Days

Why do they say that the risk of sunburn is greatest between the hours of 10 A.M. and 2 P.M.? Of course, that's when the sun is most directly overhead, but why is the overhead sun stronger? It's not any closer to us at noon, is it?

No, the ninety-three-million-mile separation between the sun and the Earth pays little attention to our lunchtime or recreational schedule. The sun is essentially the same distance from your rapidly reddening nose at all times of day. But the *strength* of the sunshine varies, for two reasons: one atmospheric and one geometric.

Picture the Earth as a sphere covered by a layer of air—the atmosphere—a couple of hundred miles thick. When the sun is directly overhead, its rays are coming down perpendicular to the atmosphere and to the ground, penetrating the least possible amount of atmosphere in the process. But when the sun is lower in the sky, its rays are coming to us obliquely and somewhat horizontally, having to penetrate much more of the atmosphere before getting to us. Because the atmosphere scatters and absorbs some sunlight, the more atmosphere the rays have to penetrate, the less intense they become. So low sun is weaker in intensity than high sun. Near sunrise or sunset, it is almost three hundred times dimmer than at noon.

But even if the Earth had no atmosphere, the sunlight would still be weaker when the sun is lower in the sky. It's a purely geometric effect of the obliqueness of the rays. The best way to see this effect is with a flashlight and an orange.

TRY IT *In a darkened room, shine the round beam of a penlight or tiny flashlight onto the surface of an orange. The penlight is the sun and the orange is the Earth. First, hold the penlight directly above the equator, in noonday position. You'll see a perfectly circular beam of sunlight landing on the Earth. Now, holding the sun the same distance from the Earth (makes you feel powerful, doesn't it?), shine the beam onto the Earth obliquely, a little to the left (west) of where you had it before, in late afternoon position. You'll see an oval-shaped light on the orange, as if the circle of sunlight has been smeared out. Well, it has been. The same amount of light is now spread out over a larger area, so of course its intensity at any one spot must be lower.*

The next time you're at the beach, notice that the black-belt suntanners use this effect to their advantage (and that

of their dermatologists' bank accounts). At any time of day, lying down makes the sunshine strike you at a somewhat oblique angle, because it is never directly overhcad, except at the equator. So what the olympic tanners do is face the sun and sit up slightly, so that the beams strike their skins as perpendicularly as possible.

NITPICKER'S CORNER:

If we wanted to, we could call this geometric effect the "cosine effect." If you work out the trigonometry, it turns out that the intensity of sunlight on the ground falls off according to the cosine of the angle between straight overhead and the sun's position. The intensity (and the cosine) decrease from full value at high noon on the equator to zero when the sun hits the horizon at sunset.

You didn't ask, but . . .

Isn't that why it is colder in the winter than in the summer?

Right on. When it is winter on the part of the Earth where you live (northern or southern hemisphere), your hemisphere is leaning away from the sun a bit. That is, the axis of the Earth wobbles, so that during winter in the northern hemisphere the North Pole is farther from the sun than the South Pole is. Because your hemisphere is leaning away from the sun, the sunshine hits its surface at a more oblique angle. The more oblique the angle, the less intense the light. And, of course, the heat. Big surprise conclusion: You're less likely to get either sunburn or heatstroke in the winter.

Mad Dogs and Englishmen

In the summertime whenever somebody wants to impress me with how hot it is, they'll say something like, "It's ninety degrees in the shade." But I can't always stay in the shade. I want to know how hot it is out in the sun, too. Is there

any way to translate in-the-shade temperatures to in-the-sun temperatures?

Afraid not. While the temperature "in the shade" is a fairly reproducible figure, the temperature "in the sun" depends too much on whose temperature you're talking about.

Different objects, including different people in different clothing, will experience different temperatures in the sun because they will absorb different amounts of different portions of the sunlight's spectrum (see p. 43). Light-colored clothing, in general, absorbs less—reflects more—of the sun's radiations than dark clothing does, so it keeps us cooler.

It's much the same with human skins: A light-skinned person may not feel as hot in the sun as a dark-skinned person will. When British imperialism was at its peak in parts of the world where the people have generally darker skins, Noël Coward immortalized that fact in his song, "Mad Dogs and Englishmen Go Out in the Mid-day Sun."

In the shade—in the absence of direct radiation from the sun—the temperature of a free object (not connected to a source or absorber of heat) depends only on the temperature of the surrounding air. That's the temperature that the weather people quote in their reports; they don't bother to say "in the shade." But in the sun, temperatures depend not only on the air's temperature but also on the absorption and reflection of heat rays by the object or person in question. These factors can vary a great deal from object to object and condition to condition.

Incidentally, there is no physical law that says that steering wheels get hotter than anything else when you park your car in the sun. It's just that the steering wheel is in a particularly sun-vulnerable position, and it's the object you need to touch most.

Green Skin and Blue Blood

Those bluish-green roofs on old churches and city halls: I understand that they're made of copper, but I've never

seen copper turn that color anywhere else. Could I make a penny green like that?

Those copper roofs have been out in the weather longer than a borrowed lawn mower—all those years that have passed since people could afford to cover roofs with that durable and beautiful red metal. Today copper is too expensive to use to shelter even the heads of politicians and priests. It is even too expensive to make pennies out of; a penny's weight of copper is now worth more than one cent. Ever since 1982 pennies have been made of zinc, with just a thin coating of copper for old times' sake. But if you really want to, you can still leave a penny out in the weather for fifty years or so, and it will turn roof-green. There's no fast and easy way to do it.

That's the reason, in fact, that copper is such a good material for covering roofs: It corrodes very slowly—much more slowly than iron rusts (see p. 81). Within a few weeks, bright, shiny copper will darken because of a thin layer of black copper oxide. Then, as the years go by, it reacts slowly with oxygen, water vapor, and carbon dioxide in the air to form the bluish-green patina that chemists identify as basic copper carbonate. In addition to roofs, this patina colors the Statue of Liberty, which is made of three hundred thick copper plates bolted together, and which has been exposed to New York City air since 1886.

Incidentally, the green color that you see on pennies in the bottoms of fountains, tossed in by people who believe that one cent will bribe the Fates into granting a wish, and that are somehow overlooked by the midnight scavengers, is not the same, chemically, as the green color you observe on the roofs. It is due to other compounds of copper such as copper chloride and copper hydroxide that don't have the same blue-green color and that don't adhere very well to the metal.

You can try to duplicate the patina of copper by buying some cheap jewelry made of brass, which is an alloy of copper and zinc. Wear an unlacquered brass ring or bracelet for a few months and the copper will react with the salt and acids in your skin to produce copper chloride and other compounds. Your skin will turn as green as Miss Liberty's. But it still won't be exactly the same shade as hers.

Many outdoor statues in pubic places are made of bronze, which is an alloy of mostly copper and tin. When the statues weather, they develop a dark-green patina similar to copper's. (The white splotches on the statues have quite a different origin.)

An interesting sidelight on copper is that instead of the red hemoglobin in human blood, which has an iron atom in its molecule, lobsters and other large crustaceans have blue blood containing hemocyanin, which is similar to hemoglobin but contains a copper atom in place of the iron. There may be some truth, after all, in the claim of revolutionaries that the world's blue-bloods are among the lowest forms of life.

You didn't ask, but . . .

What about those copper bracelets that are supposed to cure arthritis?

Nonsense. The thinking (a generous term for it) behind these voodoo baubles appears to be (1) that copper is a good conductor of electricity (which it is), (2) that there is "electrical energy in the air" (whatever that means), and (3) that a copper bracelet will therefore attract that "energy" and conduct it to your aching bones. And, of course, we all know that "energy" is good for us.

The only energy that the bracelet will generate, however, is the energy that you will have to expend in scrubbing the green stain off your wrist. (Try vinegar.)

TRY IT Take a file to a penny and you'll find out that the copper is only skin deep. Underneath, you'll see the silvery color of zinc.

Oddly enough, the penny is the only American coin that is *not* made from an alloy of copper. Nickels, dimes, quarters, and half dollars are all made from copper alloys, usually with nickel. Even nickels are only 25 percent nickel; the rest is copper.

BAR BET There is only one American coin that is not currently being made of a copper alloy. (Side bet: It's the penny.)

It's a Good Thing Air Is Transparent

How come we can see through air?

It's very simple. The molecules in air are so far apart that we're actually looking through empty space. To notice anything at all, we would have to be able to see the individual molecules, but air molecules are about a thousand times smaller than anything we can observe, even with a microscope.

We're talking about looking through pure, unpolluted air, of course. We'll get to the dirty stuff later.

Air is 99 percent nitrogen and oxygen molecules, which are roughly equal in size. The figure shows them, drawn to scale, at their normal separation distance at sea level. Notice all the absolutely empty space, nothing whatsoever in between the molecules. No wonder that light can pass through air from an object directly to our eyes, completely unhindered. And that's as good a definition of transparency as any.

But even when visible light happens to hit one of the nitrogen or oxygen molecules, it isn't absorbed. Many other kinds of molecules have the habit of absorbing light of certain specific wavelengths, or colors. When a certain specific color is absorbed out of the light, the rest of the light, lack-

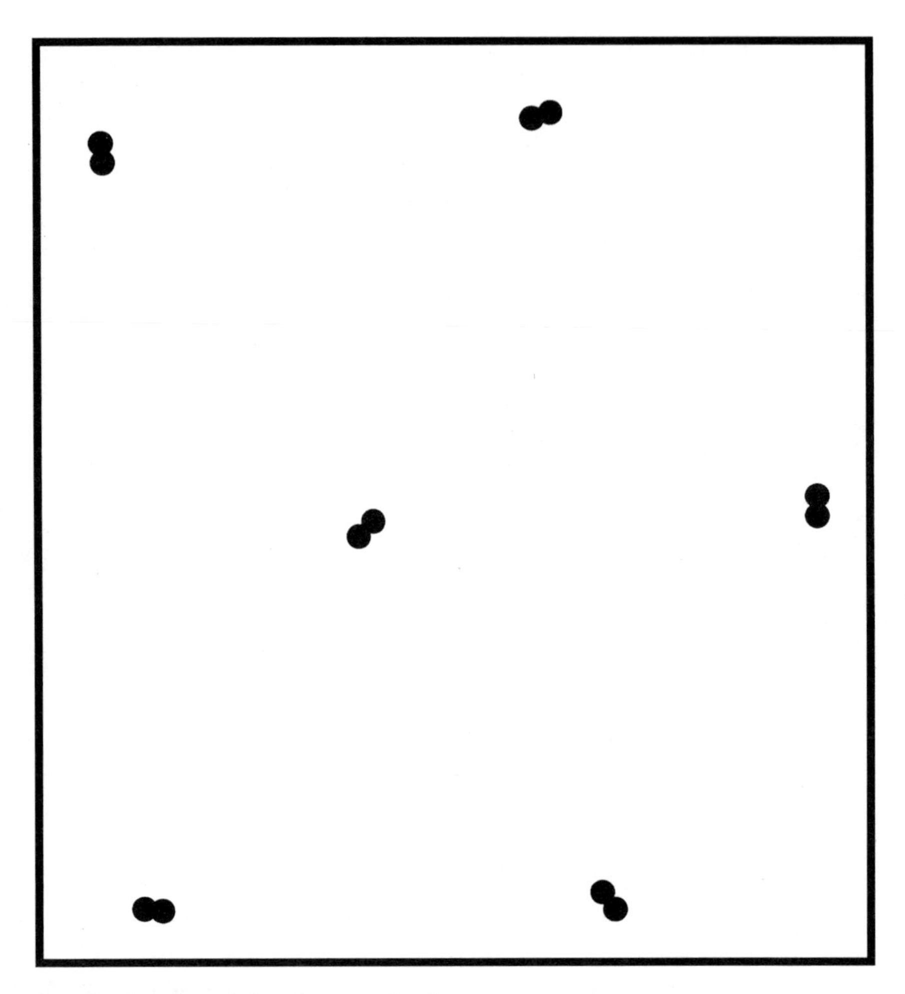

A scale drawing of the air at sea level.

ing that color, appears to us as an altered color (see p. 43). So some gases appear colored.

Chlorine gas is green, for example. If you had a glass jar full of chlorine gas, you'd still be able to see through it because the molecules are still very far apart, but the light coming to your eyes would have a greenish tinge. Transparency and color, then, are really two different things, in spite of the fact that many people call colorless plastic "clear" instead of colorless. Tinted glass is colored, but you can still see through it; it is still transparent.

Which brings us to polluted air. If you've ever flown into Los Angeles, Denver, or Mexico City, you may have seen a layer of yellowish brown sludge hanging over the city. That's air containing nitric oxide, a brown, irritating gas that is made when other nitrogen oxides from automobile exhaust react with oxygen in the air.

When pollutants, including smoke and chemical fogs, get so thick that a variety of light wavelengths are being absorbed, the air becomes generally less transparent. The molecules are still far apart, but so many of them are absorbing light—or, actually, scattering it away—that less of it gets through to our eyes. There are many locations in the "wide open spaces" where the visibility has diminished so much over a single lifetime that adults can no longer view the distant mountain peaks that they could see clearly when they were kids.

Yes, we're lucky that air is transparent. But it isn't as transparent as it used to be.

Living Under Pressure

Why do weather people talk about the barometric pressure in terms of so many inches of mercury? How can you measure pressure in inches? And what is an inch of mercury, anyway?

First of all, please don't call it "barometric pressure." The air around us has a temperature that is measured by a thermometer, a humidity that is measured by a hygrometer, and a pressure that is measured by a barometer. Television weather reporters wouldn't dream of talking about the air's "thermometric temperature" or its "hygroscopic humidity," yet they insist on referring to "barometric pressure"—probably because it makes them sound impressively scientific. Plain old "air pressure" would do just fine.

But what is this "pressure" that the air exerts? Atmospheric pressure is caused by the incessant bombardment of air molecules upon whatever they happen to be in contact with. Every time an air molecule (primarily nitrogen or oxygen) crashes into a solid or liquid surface, it exerts a force. The number of these collision forces per second on each square inch or square centimeter of the surface is a measure of the pressure. And it's not trivial; at sea level those zillions of molecule collisions add up to a total of 14.7 pounds on every square inch (1.03 kilograms per square centimeter).

Measuring pressure by counting molecular collisions is difficult. But since the atmosphere exerts a pressure on everything it's in contact with, we can use the atmospheric force upon any convenient object as our measuring standard.

In 1643 in Florence, Italy, Evangelista Torricelli decided that the atmosphere's pressure should be capable of forcing water up into an empty tube to a certain height, and that the height of that column of water would be a measure of the atmosphere's pressure. He thus invented the world's first barometer. It turned out that normal atmospheric pressure will support about thirty-four feet (about ten meters) of water. But that would make for some ridiculously large barometers.

Today we use a much heavier liquid, mercury—a silvery liquid metal—which on a typical day by the seaside can be forced only 29.92 inches or 760 millimeters up a tube.

Put another way, the atmosphere is applying the same amount of pressure that we would feel if we were submerged under 34 feet of water or under 29.92 inches of mercury.

TRY IT To get a rough idea of how much pressure the atmosphere is exerting on us, stick your toes under the leg of a kitchen or dining-room chair and put a ten-pound sack of potatoes on the seat.

That will apply a pressure of ten or fifteen pounds per square inch to your toes. That's in addition, of course, to the 14.7 pounds of pressure from the atmosphere, but you don't feel that because it is uniform all over your body. Do fish know they're under water? We're under air.

Look Not for a Silver Lining

Why are clouds white, except when they're storm clouds, and then they're black?

It's all a matter of how big the water droplets are.

That's what clouds are: collections of tiny droplets of water. The droplets are so small that under the continual bombardment of air molecules they are kept suspended in the air and do not settle out by gravity—until it rains, of course. The droplets do keep evaporating and re-forming, however, and that's why clouds keep changing their shapes.

TRY IT *On a day when there are wispy white clouds moving through a clear blue sky, lie down and watch them for a while. You'll see that they continually change their shapes as they move along with the wind. The water droplets at the edges are continually evaporating and recondensing elsewhere, which changes the outline of the cloud.*

The droplets of water in a white cloud are like tiny crystal balls. That is, they reflect and scatter light in all directions. Like water in its other forms—ice and snow—they reflect and scatter all wavelengths (colors) of light equally, so the reflected sunlight reaching us retains its full white color. (When the droplets are even smaller, smaller than the wavelengths of light, they contribute to the blue color of the sky. See p. 29.)

On the other hand, storm clouds, as you might expect, are loaded with water, just waiting for the right opportunity to spoil your picnic. The droplets of water in them are so thick that they block out light coming down from the sun and the clouds appear relatively dark against the bright sky. They're not actually black, however, any more than a shadow is.

Not Quite Cricket

I read somewhere that you can tell the temperature by listening to the crickets. How?

You count their chirps.

All cold-blooded animals perform their functions faster at higher temperatures. Just compare how fast the ants run around in cool and hot weather. Crickets are no exception. They chirp at a rate that is geared directly to the temperature. To understand their message, all you need is the translation formula.

This is not so much a biological phenomenon as it is a chemical one. All living things are governed by chemical reactions, and chemical reactions generally go faster at higher temperatures. That's because chemicals can't react with one another unless they come into contact—molecules actually bumping into molecules. The higher the temperature, the faster the molecules are moving (see p. 236) and the more often they will collide and react. Chemists like to use the rule of thumb that a chemical reaction doubles in speed for every ten-degree (Celsius) rise in temperature.

Fortunately, we warm-blooded critters maintain a constant temperature and therefore a pretty constant rate of living. Crickets, however, chirp faster when they're warmer. The best one to listen to is the snowy tree cricket of North America. But if you can't tell one cricket from another, don't worry about it. The common field cricket chirps at about the same rate.

Here's how to tell the temperature by listening to a cricket: Count the number of chirps in fifteen seconds and add forty. That will give you the temperature in degrees Fahrenheit.

When the United States finally switches to the metric system, crickets will be required by law to chirp in Celsius. You will then be able to determine the temperature in degrees Celsius by counting the number of chirps in eight seconds and adding five.

Be aware that the cricket is broadcasting the temperature where *it* happens to be. Unless you're up a tree or in the grass, your temperature is not quite cricket.

People Who Live on Glass Planets Shouldn't Burn Coal

When I entered the greenhouse at a plant nursery, I was struck by how much warmer it was than outside. Is it always warmer in a greenhouse? If so, why?

Yes, greenhouses—sometimes called hothouses or glasshouses—are always naturally warmer, without any artificial heating. But believe it or not, the main reason is *not* what everybody refers to as "the greenhouse effect."

A greenhouse is just a closed, glass container for plants. The glass lets in sunlight, which the plants need for growth, while keeping out damaging wind, hail, and animals. It also prevents the loss of moisture and keeps the humidity high, which is part of what hit you in the face when you entered. But mainly it acts as a heat valve, limiting the loss of heat from the plants to the cold, cruel world outside.

A plant, or anything else for that matter, can get cold—that is, can lose heat—in any of three ways: by conduction, by convection, and by radiation (see p. 22). Conduction isn't a problem because the leaves aren't in contact with anything, such as a mass of metal, that could conduct their heat away. That leaves convection and radiation. The greenhouse cuts down both.

Convection is the circulation of warm air or water. Because warm air rises, it can carry heat up and away from a plant leaf. Anything that prevents that warm air from escaping completely will prevent the loss of heat along with it, and any closed-in building will serve that purpose. That's the major effect of the greenhouse: It simply prevents heat loss by moving currents of air. Of course, no farmer would dream of constructing a plant-enclosure building without letting in lots of sunlight, and that's why glass-walled and glass-ceilinged enclosures were born.

A secondary effect of the glass, which nobody knew about when they invented greenhouses, is that it cuts down on heat loss by radiation. That's where the so-called greenhouse effect comes in, and here's how it works.

The photosynthesis reactions that keep plants living and growing utilize ultraviolet radiation from sunlight. After using some of the energy of this radiation, they emit lower-energy "waste radiation," infrared radiation (see p. 218), which can then be absorbed by other objects. But when an object absorbs infrared radiation, its molecules become more energetic and the object grows warmer (see p. 236). We can therefore think of the infrared radiation from the plants as if it were heat, traveling through the air in search of something to warm up.

What happens when the infrared radiation hits a glass wall or ceiling? Although glass lets in ultraviolet light pretty well, it is not completely transparent to infrared. So the glass blocks some of the infrared radiation from getting out of the greenhouse, and this trapped radiation gradually warms up everything inside.

Clearly this heating can't go on forever; greenhouses have not been known to suffer spontaneous meltdowns. After a certain point, the inevitable leakage of heat out of the house balances the infrared buildup inside, and the temperature levels off at a moderately warm level—warmer than if the glass were completely transparent to infrared radiation.

You didn't ask, but . . .

What is "the greenhouse effect," in connection with global warming?

It is the effect of infrared radiation-trapping by the Earth's atmosphere, which can raise the average temperature at the surface of the entire globe, just as the trapping of infrared radiation within a greenhouse raises the temperature inside.

The overall temperature of the Earth's surface—averaged over all seasons and climates—depends on a fine balance between the amount of the sun's radiation that comes down to us and the amount that is reflected or reradiated out into space. About one-third of the sun's energy that hits the Earth is reflected back out; the rest is absorbed by clouds, land, sea, and sunbathers. Most of the absorbed energy soon degenerates into heat, or infrared radiation, just as it does in the plants in the greenhouse.

Hanging over the Earth's radiating surface is a transparent canopy, similar to the glass in a greenhouse. It isn't glass, however. It's a layer of air—the atmosphere. Like glass, the Earth's atmosphere is quite transparent to most of the sun's incoming radiations. But certain gases in the atmosphere, mainly carbon dioxide and water vapor, are very efficient absorbers of infrared radiation. Just as the glass does in the greenhouse, these gases block the escape of some of the infrared radiation, trapping it down here near the surface. The Earth is thus somewhat warmer than it would be if there weren't any carbon dioxide and water vapor in the atmosphere.

The Earth's radiation-in and radiation-out processes are delicately balanced, so that our planet has stayed at pretty much the same average temperature over thousands of years. But recent human activities have been changing that. Since the industrial revolution began about a hundred years ago, we have been burning coal, natural gas, and petroleum products at an ever-increasing rate. When these fuels are burned, they put carbon dioxide into the air (see p. 106). The amount of carbon dioxide in the atmosphere has therefore risen about 30 percent in the past hundred years. More carbon dioxide means more earthbound infrared radiation and higher temperatures.

The amount of global warming that can be caused by a given amount of carbon dioxide in the atmosphere is hard to estimate. On one hand, the oceans and forests diminish the effect by soaking up carbon dioxide from the air. On the other hand, the world's huge rain forests are rapidly being devoured by logging and by burning, which compounds the problem by putting even more carbon dioxide into the air. Even though we may not be able to determine the specific amount of global warming caused by man-made carbon dioxide, it has been pretty well established that the Earth's average temperature *has* been rising unnaturally during the past hundred years, and that it may rise by 1.5 to 4.5 degrees Celsius (0.8 to 2.5 degrees Fahrenheit) over the next hundred years as the amount of carbon dioxide in the air doubles.

A temperature increase of only a few degrees could have catastrophic consequences. Slightly warmer Arctic and Antarctic climates would melt huge amounts of ice, raising the level of the oceans and inundating coastal cities all over the world. At the very least, there would be changes in global weather patterns, with significant consequences in food production and water supplies.

Our planet's atmospheric greenhouse is apparently just as fragile as if it were actually made of glass.

Ridiculous? No. Sublime

When there is snow on the ground, I've noticed that it slowly melts away over a period of a week or two, even when the temperature stays well below freezing. Where does it go?

The snow isn't melting; it is actually going straight off into the air as water vapor, without having to melt into liquid water first.

We might be tempted to say that the snow is evaporating, but scientists prefer to reserve the word "evaporation" for liquids only. So when a solid "evaporates" they call the process *sublimation*. In our everyday experience, we rarely notice solids *subliming* because sublimation is generally a much slower process than the evaporation of liquids.

Here's how sublimation takes place. The molecules at the surface of a chunk of solid are not attached as firmly as are the molecules within the bulk of the piece. While the molecules in the bulk are bonded to their brethren in all directions, top, bottom and all around, the surface molecules are bonded in every direction but their "tops," which are exposed to the great outdoors. They are missing a bit of adhesion to the rest of the solid.

If you consider that molecules are always jiggling around to some extent (see p. 236), it is not too hard to imagine

that an occasional surface molecule might break loose and fly off into the air. That molecule has sublimed. The molecules of liquids are more loosely tied together than are the molecules of solids, so the probability that a liquid molecule will break away is much greater. That's why liquids generally evaporate much faster than solids sublime.

Snow is a great candidate for sublimation because it is made of intricate, lacy crystals with large surface areas; and the more surface molecules there are, the more molecules can sublime. But you can even see solid chunks of ice sublime. Ever notice how old ice cubes shrink in the freezer?

Different solids have different tendencies to sublime because they are made of different atoms or molecules that are tied together with different strengths. Fortunately, the atoms of metals are tied together very tightly, so gold and silver do not evaporate at all. On the other hand, the molecules of some organic solids are tied together rather loosely, so they have a substantial tendency to fly off as vapor. Moth crystals and deodorizing cakes are usually made of paradichlorobenzene, an organic solid that is a sublime sublimator, so to speak. Its strong-smelling vapor quickly fills the air and kills both moths and our ability to smell foul odors.

TRY IT Measure the length of a convenient icicle during a cold spell. Then come back in a day or two and measure it again. Make certain that the temperature hasn't gotten above freezing in the meantime, so that there hasn't been any melting. You will see that the icicle has become smaller by sublimation.

You didn't ask, but . . .

How do they make freeze-dried coffee?

By the sublimation of ice.

Freeze-dried coffee differs from ordinary instant coffee in an important way. To make either kind of fast-beverage powder, they first brew two-thousand-pound batches of incredibly strong coffee. If they are making instant coffee, they then quick-dry this thick brew by dropping it down through a high-temperature chamber. All the water evaporates and only powdered solids reach the bottom. Unfortunately, the heat drives off some of the most flavorful coffee chemicals.

On the other hand, when making freeze-dried coffee, they freeze the strong brew into blocks of solid coffee-ice. Then they pulverize it into granules and put it in a vacuum chamber, where the water molecules are sucked directly out of the ice—by sublimation. Most connoisseurs of coffee-in-a-hurry believe that, compared with ordinary instant coffee, freeze-dried coffee has a truly sublime flavor.

Sizzlin' Snowflakes, Batman, Why Is It Getting so Warm Around Here?

This is going to sound crazy, but I swear it's true. I spend a lot of time outdoors in the winter, and every time it begins

to snow, I've noticed that the air becomes warmer! You'd think that in order to start snowing, the air has to get colder, not warmer. What's going on?

You're a good observer. It really does get warmer when the snow begins to fall.

Think of it this way: In order to *melt* a lot of ice or snow, you have to add heat to it. So when a lot of water *freezes* into ice or snow, which is the reverse process, that same amount of heat has to come back out again. It does, and it heats up the air. The question is *why* that heat comes out.

First of all, water in the air isn't going to freeze into snowflakes at all unless the temperature is lower than 32 degrees Fahrenheit or 0 degrees Celsius. You've never heard a weatherman predict "temperatures in the mid-seventies with occasional snow flurries." So all the necessary cooling—or temperature lowering—that you rightfully expect will already have taken place by the time the first flake forms. Nothing the least bit remarkable there.

As soon as the water begins to freeze into snow, though, something new begins to happen. In a droplet of liquid water, the molecules are quite loose; they're sliding around each other freely and randomly. But when that droplet freezes into the beautifully-shaped ice crystal we call a snowflake, the water molecules must snap into a rigid crystalline formation (see p. 202). They have less energy in the rigid snowflake formation than they had in the chaotic liquid form. It's like a schoolteacher taming a bunch of wild kids by making them line up in the hallway. If the water molecules now have less energy in the snowflake than they had in the water droplet, the excess energy had to go somewhere. It did. It went off into the air as heat.

For each gram of water that freezes into a gram of ice or snow (a gram of snow would make a snowball about the size of a marble), 80 calories of heat are released. If it stayed in that gram of water, that amount of heat would be enough to raise its temperature from freezing to 80 degrees Celsius or

176 degrees Fahrenheit! But, of course, the heat doesn't stay there, or the water would never freeze. It is swept away into the surrounding cold air.

Thus, when any given gram of water turns into snowflakes, the local surrounding air gets a gift of 80 calories of heat. Multiply this by the zillions of grams of water that are freezing at the beginning of a snowfall, and it's no wonder you feel warmer.

You didn't ask, but . . .

When frost threatens, why do people spray their tomato plants with water to protect them from freezing?

The water on the wet plant leaves will begin to freeze first, releasing its 80 calories of heat per gram. The leaves will absorb this heat and stay warmer than they would otherwise have been. Gardening books are wrong when they tell you that the frozen water protects the leaves by acting as an insulator. The insulation value of a thin coating of ice is nil.

BAR BET *A snowfall makes the weather warmer.*

It's All in the S'Know-How

As a skier, I often have to settle for artificial snow, made by those snow-making machines. Do they just pump a spray of water into the air and let it freeze?

No. That wouldn't work very well, except perhaps in extremely cold weather. And by the way, the machines don't produce actual snowflakes; they make tiny beads of ice, each one around ten thousandths of an inch in diameter.

The simple spraying of water wouldn't work because when water freezes, it gives off quite a bit of heat (see p. 163). That happens because when water molecules transform themselves from a liquid to a solid, they have to stop moving around and settle down into rigid positions, and the movement energy that they previously had must wind up someplace. If large amounts of sprayed water were to be frozen fairly near the ground, the liberated heat would warm up the air substantially and defeat much of the whole purpose; the ersatz snow would be wet and not very cold.

When real snow forms in nature, on the other hand, the heat is given off way up in the air someplace where the snow was created, and it doesn't significantly warm up those beloved slopes. That's why at many ski resorts the snowmaking machines do their spraying from high towers, letting the wind carry off the heat.

In any event, some extra cooling is necessary to counteract the heat that is released on freezing. The machines accomplish this by spraying not just water, but a mixture of water and high-pressure air, at around 118 pounds per square inch. When compressed air, or any gas for that matter, is allowed to expand suddenly, it gets cold. By pushing aside the atmosphere or anything else that it's expanding against, gases use up some of their energy (see p. 135). The coldness of the expanding air more than makes up for the warming that comes from the freezing water. And then for good measure, the blasted-out water droplets are cooled further by evaporation (see p. 177).

Strangely, though, no matter how cold the water gets, it won't spontaneously freeze. Everybody says that water freezes at 32 degrees Fahrenheit (0 degrees Celsius), but they should add, "provided that something stimulates it to begin the freezing process." Water molecules can't begin to settle into the highly specific orientations and rigid positions that they must have in an ice crystal without some kind of "starting gun" to shake them into place.

It's a fact that water can be cooled far below its normal freezing temperature—that is, it can be substantially *supercooled*—without freezing. It is much too tricky to try at home, but under careful laboratory conditions, pure water can be supercooled down to 40 degrees below zero without freezing. (Fahrenheit or Celsius, it doesn't matter; 40 below happens to be 40 below on both scales. See p. 242).

A mechanical shock can shake up the molecules in supercooled water droplets enough to make them fall into their assigned places in an ice crystal. In the case of the snow-making machine, the shock is supplied by the high pressure air blast, which shoots the microscopic droplets of water out of nozzles at near-sonic speeds.

An interesting new wrinkle in snow-making machines is the addition of a certain species of harmless bacteria to the sprayed water-air mixture. It has been found that these bacteria, which live on plant leaves almost everywhere, help water to freeze more quickly. In doing so, they may be helping to save the plants from frost damage (see p. 165). They apparently perform the same quick-freeze service in the snow-making machine, producing a greater yield of artificial snow by freezing more of the tiny water droplets before they have a chance to evaporate.

Snowball Fight!

I've been having an argument with a friend about what holds snowballs together. He said that because snowflakes are jagged, they must hook together like Velcro. I wasn't convinced. Was he right?

It's a nice idea, because snowflakes certainly do have beautifully complex shapes, with spikes, lacy edges, and all the rest. But interlocking hooks and loops are a bit too much to expect. Besides, they're much too fragile and brittle; when you pack them together they suffer a crushing experience.

The answer lies in the fact that pressure can melt frozen water—ice or snow (see p. 211). When you press the snow together tightly, the pressure melts certain portions of the flakes. They can then slide over one another on the resulting water film, and the ball compacts. But the main body of the snow is still below the freezing temperature, and the melted parts quickly refreeze. This refrozen ice acts like a cement that holds the whole thing together.

If you are intrepid enough to be making snowballs with your bare hands, your body heat is also melting a thin layer on the outside surface. When this layer refreezes, you've got yourself a case-hardened weapon. Although the Geneva Convention strictly forbids it, some combatants dip their snowballs in water to make them even harder.

TRY IT For Yankees only: Put a dark-colored dish in the freezer and wait for snow. When it starts snowing (that's usually when the flakes are biggest), take the dish outside with a magnifying glass—the most powerful one you can find. A cold microscope and slide would be even better. Catch some snowflakes on the dish or slide and quickly examine them with the magnifier. What beautiful crystals! If the snow catches you unprepared, a piece of cold, dark cloth will also work as a flake catcher.

You didn't ask, but . . .

Does it ever get too cold to make snowballs?

Yes. Every northern kid knows that wet snow makes the best snowballs. That's because snow that's not too much colder than the freezing point is easy to pressure-melt, and it will therefore compact into an effective projectile. But when the snow is too cold, the strength of even the most belligerent bully will be inadequate to pressure-melt and refreeze many flakes, and the snow will fall apart into useless shrapnel.

Things That Go Boomp in the Night

How do they make all those colors in fireworks?

They add chemicals to the explosive mixtures that emit specific colors of light when subjected to heat. You could throw some of these same chemicals into your fireplace if you thought that a green fire, for example, might be more romantic.

When you throw an atom into a fire, it can pick up some of the fire's energy by making its electrons move faster. These "hot" electrons are just dying to return to their relatively sluggish, natural energy states (Techspeak: their *ground states*). The easiest way for them to do that—easy for an electron, that is—is to discharge their excess energy in the form of a blast of light. When enough atoms in a fire are simultaneously taking on heat energy and throwing it back off in the form of light, we can observe a very bright light.

Every type of atom or molecule has a unique set of electron energies to begin with. Therefore, each type of atom or molecule in the flame will be able to take on and throw off only its own unique amounts of energy. That is, different atoms and molecules will be emitting different wavelengths or colors of light. (Techspeak: every atom or molecule has its own, unique *emission spectrum*.) Unfortunately for the fireworks manufacturers, most atoms and molecules emit their light in colors that humans can't see: in the ultraviolet or infrared regions of the spectrum. But the atoms of some elements emit light in brilliant colors that we can see.

Here are some of the kinds of atoms (in the form of their chemical compounds) that are used to make the colors in fireworks: Reds: strontium (used most often) makes a crimson light, calcium makes yellowish red, lithium makes carmine. Yellows: sodium makes a bright, pure yellow. Greens: barium (used most often) makes yellowish green, copper makes emerald green, tellurium makes grass-green, thallium makes bluegrass green, zinc makes whitish green. Blues: copper (used most often) makes azure, arsenic makes light blue, lead makes light blue, selenium makes light blue. Violets: cesium makes bluish purple, potassium makes reddish purple, and rubidium makes violet.

TRY IT The next time you have a fire going in the fireplace or on the beach, sprinkle some crushed table salt or some powdered bicarbonate of soda on it and you'll see the brilliant yellow flame color that sodium makes. If you have one of those salt substitutes around—the kind that's sold for people on sodium-free diets—toss some of it onto the fire. It contains potassium chloride instead of sodium chloride, and you'll see potassium's characteristic reddish-purple flame color. If you happen to be taking lithium for a manic-depressive condition, your medicine will make the most beautiful red flames you've ever seen.

You didn't ask, but . . .

How do they make all those colors in neon signs? Is it just colored glass?

No, the colors are actually glowing atoms, stimulated by electricity. It's pretty much the same as making the colors in fireworks: Stimulate atoms with energy, and they'll quickly get rid of the excess energy by emitting light of their own characteristic colors.

There are a couple of differences (fortunately) between fireworks and neon signs. In neon signs, the atoms are in the form of gases inside of glass tubes that are shaped to spell out words or make pictures. Instead of explosions, the gas atoms are stimulated by a high-voltage electric current passing through the tube from one end to the other. If the gas happens to be neon, it emits that familiar orange-red color that announces the presence of NICK'S BAR AND GRILL.

Other gases give off their own colors of light when excited by an electric current. For example, helium makes a pink-violet light, argon makes bluish-purple, krypton makes a pale violet, and xenon makes blue-green. Other colors are made by mixing gases or by coating the insides of the tubes with solid materials that glow with their own colors.

Nobody has yet been able stop people from calling all of these signs "neon," regardless of what gas happens to be inside the tubes.

BAR BET That blue "neon" beer sign has no neon in it at all.

Up, Up, and a . . . Why?

What eventually happens to a helium-filled balloon when you let it go outdoors? And why do helium balloons fall up, anyway? Doesn't gravity act on helium, as well as on everything else? If something is moving upward, there must be an upward-pushing force, mustn't there? So what's the force? Antigravity?

Antigravity? We don't use that word in this book. Science fiction is two shelves over to the left.

Surprisingly, there is no *upward*-pushing force. It's just that there is less *downward*-pulling force on the helium than there is on the air that surrounds it because helium gas is lighter than an equal volume of air (see p. 243). Gravity pulls less strongly on the lighter helium atoms than it does on the heavier air molecules. The air will therefore tend to move down past the helium, or—same thing—the helium will be observed to move upward past the air. If you were inside the helium balloon, you might be wondering, Why is all that air rushing downward past me?

When you release a piece of wood under water, you're not surprised to see it zoom upward through the water, are you? That's because wood and water are such familiar materials that you expect wood to float on water (see p. 182).

Helium and air, however, being gases instead of solids or liquids, are not as familiar to us; we can't see them, pour them, grab them, or throw them. But they are *matter* (substance) all the same, made up of tiny particles that are tugged upon by the Earth's gravitational field, and they respond in the same way as solids and liquids do. The force of gravity is proportional to the mass of the particles, be they in the solid, liquid, or gaseous form.

When you let go of a helium balloon outdoors, several things happen. As it ascends, it encounters changing conditions of both air pressure and air temperature. As far as the pressure is concerned, it decreases pretty regularly as the altitude increases. That's because the atmosphere is a layer of air enveloping the earth, held down tightly against the globe by gravity. The higher you ascend into this layer, the less of it remains above you pressing downward, so you feel less air pressure. And so does the balloon.

At any given time, a rubber balloon is a certain size because the outward-pushing pressure of the gas inside is counteracted by the inward-pushing pressure of the atmosphere outside (plus, of course, the inward-contracting tendency of the rubber). When the atmospheric pressure decreases, the outward-expanding tendency of the helium gas can prevail, and the balloon will expand. So as the altitude increases, the balloon tends to get bigger. Hold that thought.

Now, what are the effects of decreasing temperature? We know that all gases will try to expand when heated and contract when cooled. That's because the molecules of a hot gas are bouncing around faster and pushing harder against any walls that are attempting to contain them. Our particular container of helium is ascending into colder and colder air; the average temperature of the earth's atmosphere decreases from about 65 degrees Fahrenheit (18 degrees Celsius) at sea level to about 60 degrees below zero Fahrenheit (–51 Celsius), at an altitude of 6 miles (10 kilometers). So as the balloon rises and gets colder, it will tend to shrink.

We now have two counteracting tendencies: an expansion due to the atmospheric pressure decrease and a contraction due to the atmospheric temperature decrease. Which tendency will win out?

The rules that govern the expansion and contraction of gases are well-known; scientists lump them into a mathematical equation called the *gas law*. Using this equation, they can actually calculate the effects of varying pressures and temperatures on a gas. If you do the calculations for our rising helium balloon (and I did), you find that the expansion due to the pressure decrease is a much bigger factor than the contraction due to the cooling.

So the net effect on the balloon is that it gets bigger and bigger as it rises until—*pop*!—the rubber will stretch no farther and it bursts, eventually fluttering down into somebody's picnic mustard. The helium gas, now unfettered, just keeps rising through the atmosphere until it reaches a level where the air is so thin that a balloonful of it would be just as light as a balloonful of helium, and that's where it will remain until doomsday.

NITPICKER'S CORNER:

Well, not exactly until doomsday. Because of winds, weather, and other mixing phenomena, we can always find a little helium in the air at any altitude—on the average, about five helium atoms for every million air molecules. And at the top of the atmosphere, some of them even escape from the Earth entirely.

Moreover, we must admit that other happenings can interfere with our rather neat picture. Our balloon may not even get high enough to explode, because the amount of helium in it isn't enough to carry its payload of rubber high enough, and it'll settle out at a maximum altitude. Then, winds could blow it about for days, until enough helium has seeped out (helium atoms are extremely tiny as particles go, and can diffuse right through the rubber) that the weight of the rubber brings it down. You've probably seen that happen to a balloon left on your ceiling for a couple of days.

And by the way, many helium balloons these days are

made out of aluminized Mylar—a tough plastic film coated with a very thin layer of aluminum—rather than rubber. They will last a lot longer and go a lot higher before meeting their fate. Commercial jet airplanes have been known to spot them miles high, speeding along in the jet stream.

Look! Up in the Sky! It's a Bird. It's a Plane. It's the Goodyear Prune!

Blimps, airships, dirigibles, lighter-than-air-craft—whatever you call them: They're filled with helium gas, right? But when they get heated and cooled by the sun and the weather, the gas has to expand and contract, doesn't it? How do they take care of that? Does the whole balloon expand and contract?

No, that would knock the sponsor's neon signs off the sides, and that would never do because today's blimps are nothing but flying billboards. Instead, they use a clever system of swapping helium and air back and forth.

The blimp is, as you've noted, essentially a big rubber bag full of helium. The contraption floats in the air because the whole thing—helium, rubber bag, gondola, engine, crew, and joyriding local politicians—together weigh less than an equal volume of air (see p. 182).

On a hot day with the sun beating down on it, there can be quite a pressure buildup, tending to make the helium expand. But they can't just vent all that expensive helium out into the air. Moreover, what would they do when the bag cools and they need more helium to keep it from looking like a flying prune?

There is a small, separate bag of air inside the big bag of helium, like an air balloon inside a helium balloon. They are arranged so that when the helium expands, it just pushes some cheap, old air out of the ship. And when the helium contracts, they make up for the shrinkage by blowing more air into the inner bag. Or else they get the politicians to make speeches into it.

Why Astronauts Get Such a Warm Reception

Outdoors, the stronger the wind blows, the colder I feel. I think I understand that. But when a returning space shuttle plunges into the atmosphere, the passing air heats it up so much that they have to protect it from burning up like a meteorite, even though the air is a lot colder up there. How come, when the "wind" is strong enough, it turns from a cooling wind into a burning wind?

First of all, when it is very windy, the cooling effect on your skin has little to do with the evaporation of perspiration, in case that's what you were thinking. That effect (see p. 177) peters out as soon as there is enough wind so that all the perspiration has already evaporated. A strong wind cools us because the moving stream of air molecules carries off heat from our bodies; the faster it passes by, the faster it takes away heat. The moment your skin heats up an adjacent air molecule, it is whisked away, carrying your hard-earned body heat along with it. Clothing protects you largely because it keeps those thieving air molecules from skimming alongside your skin.

As to the space shuttle: The first thing you have to do is to forget about *friction*, the word that newspapers and magazines invariably use to "explain" the heat of atmospheric reentry. Friction comes from the rubbing of two solids together. For gases, the word is meaningless. The molecules of a gas are so far apart, with so much empty space between them, that a gas is thoroughly impotent at "rubbing" against anything. The only thing that gas molecules can do is to fly around and collide randomly with objects, like a horde of house flies dashing themselves madly against the display cases in a manure museum. (Sorry about that, but it describes exactly how gas molecules behave.)

The air is much colder and thinner at about forty miles up, where the heating of a reentry vehicle really begins to get serious. But when the "wind" is whooshing past the shuttle at around eighteen thousand miles per hour, which is the speed at which the vehicle enters the atmosphere, we have quite a different situation from earthbound zephyrs. At eighteen thousand miles per hour, the shuttle is actually moving much faster than the individual air molecules are as they flit randomly about. The average flitting-about speed of molecules is essentially their temperature (see p. 236).

The result is exactly the same as if the shuttle were standing still and the air molecules were bombarding it at their regular speed *plus* eighteen thousand miles per hour. That makes a total molecular speed that is equivalent to a temperature of several thousand degrees. The shuttle, then, feels as if it were being exposed to air at a temperature of several thousand degrees. If it hadn't been covered with a highly heat-resistant ceramic material that uses up energy by melting off, it would indeed burn up like a meteorite. (And yes, that's why meteorites burn up.)

Even ceramics, however, cannot long endure at such high temperatures. Fortunately, the leading edges of the space shuttle are preceded by a shock wave—a layer of air molecules that pile up because they simply can't get out of the way fast enough. This layer of air acts as a front bumper on the vehicle. It soaks up the brunt of the heat energy by breaking down into a glowing cloud of atomic fragments and electrons—what scientists call a *plasma*. That's what makes the V-shaped "prow wave" that you see in those telephoto pictures on TV.

6

Water, Water Everywhere

Water is the most abundant chemical compound on Earth. It covers about 75 percent of our planet's surface, making the Earth look blue and white from space. (The white clouds, of course, are also water.) The total amount of water on Earth, including oceans, lakes, rivers, clouds, polar ice, and chicken soup, amounts to one-and-a-half-billion billion tons. In fact, we ourselves are more than half water: A typical 150-pound male is about 60 percent water; females average closer to 50 percent; fatter people have a lower percentage. Babies are as high as 85 percent water, not counting the diapers.

Water possesses some of the most unusual properties of any chemical in the universe, and yet it is so familiar to us that we take it completely for granted. But what is really happening when we boil it, freeze it, float on it, or perspire it? In this section, we'll peek beneath the surface of our everyday encounters with this most remarkable liquid.

How Sweat It Is!

I know that people sweat as a mechanism for keeping cool, because when perspiration evaporates it has a cooling

effect. But why is evaporation a cooling process? Just because a liquid is evaporating, why should its temperature go down? Or does it?

Sorry, but the answer has to be, "It does and it doesn't." Maybe that's why people go around simply parroting the prefabricated, though hardly enlightening, answer—that "evaporation is a cooling process."

We notice that our sweat glands are exuding a liquid—water containing a little salt and urea—onto our skin only at certain times, such as (a) when we're hot, (b) when we're exerting ourselves strenuously, or (c) when we're about to deliver a speech and can't find our notes.

In reality, however, our perspiration process is always working, even in cold weather. It's an essential mechanism for keeping our body temperature constant. In situations like a, b and c above, the perspiration is being generated faster than it can evaporate, so we notice the buildup of actual moisture on our skin.

Dogs, being called upon much less frequently to deliver speeches, do not have sweat glands on their skin (except, oddly, on the pads of their feet). So they hang out their extraordinarily long tongues and pant, which hastens the evaporation of saliva, thereby cooling the air supply to their lungs.Other animals sweat to varying degrees. Pigs do indeed "sweat like pigs" on occasion, although they also like to cool off by wallowing in mud, as do elephants and hippopotamuses. Not much different, actually, from our own habit of taking a quick dip in the pool.

But what, exactly, is evaporation? It is the process in which certain molecules at the surface of a liquid simply decide to depart from their brethren and fly off. As more and more molecules leave, the amount of remaining liquid diminishes. You've seen it happen dozens of times: Wet floors dry up and laundry dries on the clothesline.

If we want to hasten evaporation, we can do two things: heat and blow. Heating the liquid gives more of its molecules the energy they need to escape. Hence we have hair driers and those abominable hot-air hand driers in public rest rooms. Their blowing disperses the crowd of just-evaporated water molecules and makes room in the air for more. Blowing on hot soup to cool it off is a classic, though inelegant, application of this principle. Another example: You'll feel cold coming out of the bath if the room is drafty, even though the air temperature may be quite comfortable.

TRY IT Blow on the back of your hand and it'll feel cool, even though your breath is warm and you may think you're not perspiring.

Blowing speeds up the evaporation of the small amount of moisture that's always on your skin. Outdoors, you'll always feel colder when it's windy. The "wind chill factor" that northern weather broadcasters love to frighten us with during the winter is an attempt to take this phenomenon into account. Unfortunately, it only applies when you're naked.

Okay, so why should the exodus of water molecules lower the temperature of the remaining liquid, and therefore lower the temperature of whatever the liquid is in contact with? This may sound almost spooky, but the process of evaporation is highly selective. It preferentially picks out and removes the faster (hotter) molecules, leaving the cooler (slower) ones behind. Here's how.

The molecules of any liquid are in constant motion: sliding around, jiggling back and forth, darting about, colliding with one another, and generally acting like a bowlful of ants. The higher the temperature, the faster the molecular motion (and the faster ants move, if you really want to know). In fact, that's what temperature *is*: a measure of the average kinetic energy (energy of motion) of all the molecules in the substance.

The important word here is *average*, because at any given

temperature, the molecules are by no means all moving at the same speed. Some may be moving very fast because they've just been kicked by a collision with another molecule. Meanwhile, the molecules that kicked them are moving more slowly because they have just given some of their energy to the molecules they hit. Go to the nearest pool table and you'll see that the cue ball slows down substantially when it hits another ball, while the struck ball goes sailing away at high speed. But the *average* energy of the two balls—their "temperature"—remains the same.

Now, at the surface of a liquid, which molecules do you suppose are most likely to leap into the air and evaporate? The highest-energy ones, of course. And that will lower the average energy, the temperature, of the molecules that are left behind. Thus, as a liquid evaporates, it cools down.

But that's not the end of the story. The cooling can't go on without limit. Did you ever see an evaporating puddle spontaneously freeze itself solid? No, what happens is that as soon as the evaporating liquid begins to cool a bit, heat flows in from its surroundings and replenishes the population of high-energy molecules, which in effect keeps the temperature constant.

"Aha!" you say. "Then we're back to square one. If the evaporating liquid never gets a chance to stay cold, why does evaporating sweat cool me?"

Well, where do you think that replenishment heat has to come from? Your skin. As the evaporation proceeds, then, the sweat layer itself never does get a chance to cool down very much because it keeps taking up heat from your skin and throwing it off into the air in the form of its hottest molecules. The sweat is just a go-between, helping your skin to throw heat away.

The rate at which liquids evaporate depends on how tightly they are bound together in the liquid. In a liquid where the molecules aren't very strongly attached to each other, they can leave the crowd more easily and the liquid evaporates more rapidly. Some liquids evaporate so fast—are so volatile—that replenishment of heat from the surroundings can't keep up. In that case, the temperature of the liquid really does go down.

Ethyl alcohol is one of those volatile liquids. It evaporates more than twice as fast as water.

TRY IT Put some alcohol on your skin (isopropyl, or rubbing, alcohol will do) and you'll feel a much greater cooling effect than you get with water.

That's because "hot" alcohol molecules are leaving at such a rapid rate that they out-pace your body's ability to warm the area back up to body temperature.

Ethyl chloride is an extremely volatile liquid whose molecules don't really want to have much to do with each other and are just dying to leave home. It evaporates about a hundred times faster than water. Put some ethyl chloride on

your skin and it'll get so cold that it numbs your sensations. Doctors use it as a local anesthetic for minor skin surgery.

Archimedes Unprincipled

How can a hundred-thousand-ton aircraft carrier possibly float on water? I know that if it were a solid chunk of steel it would sink, and that it isn't solid; it's hollow. But how does the water underneath know that?

The pat answer to the everyday puzzle of why things float invariably goes like this: "According to Archimedes' principle, a body immersed in a fluid is buoyed up by a force equal to the weight of the fluid displaced. And that's why things float." Perfectly correct, of course, but just about as illuminating as a firefly wearing an overcoat.

Obviously the water underneath a ship has no information as to whether the object pressing upon its surface is a solid lump or is a sea-going Swiss cheese (except for holes in the hull, which we'll get to). Nevertheless, most of our experience with floating things, from dugout canoes to plastic foam, makes us believe that hollowness—air spaces in the interior of an object—is somehow necessary. It is not; hollowing things out is just a way of making them lighter. Light things float and heavy things sink. Which is just what you would have expected if that old Greek Archimedes hadn't muddied the waters, so to speak.

The question is, just how light does an object have to be in order to float? And the answer is, lighter than an equal bulk or volume of water. The weight of a given volume of a substance is called its *density*. Density is usually expressed as the number of pounds per cubic foot of the substance or the number of grams per cubic centimeter. If an entire ship, considered as a huge, complex conglomeration of metal, wood, plastic, air spaces, and so on, weighs less than an equal volume of water—that is, if the ship's *density* is less than the density of water—then it will float. A block of wood floats because its density is only about six-tenths as much as the density of water, so no hollowing out is needed.

If we want to float a hundred thousand tons of aircraft carrier, then, we'd better do some serious hollowing out to get its overall density down. That's no problem, of course, because it gives us quite a few convenient places to stow such necessities as airplanes and sailors.

To find out why a floating object has to be less dense than water, let's do a little experiment. Let's lower the one-hundred-thousand-ton aircraft carrier *Admiral Nimitz* (the world's largest) very gently into a rather large bathtub of water big enough to float the ship. Gravity does the lowering job for us by pulling the ship downward into the water with a force equal to its weight. (That's what weight *is*.) But as it enters the water, the ship makes a hole in the water. That is, it must push some water aside and upward against water's natural gravitational preference for settling down (see p. 205). So as gravity pulls the ship down, some water is forced up against gravity. Notice the level rising in the bathtub?

How much water can eventually be lifted *upward* against gravity? Only as much weight as the *downward* pull of gravity on the ship. In other words, the weight of the water that is lifted or displaced will be equal to the weight of the ship. When that limit is reached—one hundred thousand tons of displaced water in the case of the *Nimitz*—the ship stops settling down. By God, it's floating!

But notice that each cubic foot of displaced water must have been displaced by exactly one cubic foot of the ship's volume. That means that the volume of ship below the water line is the same as the volume of one hundred thousand tons of water. But because the water is more dense than the ship, one hundred thousand tons of water take up less space than one hundred thousand tons of ship—less than the entire ship's worth. So the amount of ship that is below the water line is less than the whole ship. Which is fortunate, because that means that the water line is only part way up the hull, where sailors invariably prefer it to be. All because the ship's overall density has been made to be less than that of water.

You didn't ask, but . . .

How about submarines? They prefer to float sometimes and to sink sometimes. How do they change their buoyancy?

Very simply. They change their amount of internal air space, thereby changing their density. You want to dive? You let water into your ballast tanks. You want to surface? You blow the water out with compressed air. It gets a bit tricky in reality, though, because the density of seawater actually varies a bit, depending on depth, temperature, and salinity (saltiness). The density of the submarine, therefore, has to be continually adjusted.

TRY IT Seawater is about 3 percent denser than fresh water. A ship in ocean water is therefore buoyed

up by a 3 percent greater force than a ship in a lake, and it therefore floats a bit higher. The Dead Sea and the Great Salt Lake are so dense from their high salt contents that their buoyancy is astonishing. Try floating in one of them if you ever get the chance. You'll only sink in a few inches. It's an amazing sensation.

You didn't ask this either, but . . .

According to Archimedes, there is some kind of buoyant force that pushes upward against any object that is placed in water. Where does that force come from?

If you doubt that the water exerts an upward pressure, try to submerge a balloon in the bathtub. You'll feel a substantial upward push that resists your downward push.

When we lowered the *Admiral Nimitz* into our giant bathtub, the water level rose; it got deeper. As every diver knows, deeper water means higher pressure. This increased pressure is present everywhere throughout the water in the bathtub, because water cannot cushion or absorb a force, the way a spring or a piece of rubber can. The water must transmit its increased pressure in all directions to everything it is in contact with, including the ship's hull. All the north-east-south-west horizontal pushes on the hull cancel each other out, leaving only an uncanceled upward push. This is the pressure that pushes the ship upward against gravity. Voilà! Buoyancy.

Okay, I know what you're thinking. Aircraft carriers operate much more frequently in oceans than in bathtubs. Am I telling you that the level of the ocean was raised when the *Nimitz* was launched? I certainly am. Spread that one hundred thousand tons of water over the surface of the entire Atlantic Ocean, though, and it comes to a pretty meager rise, quite unlikely to flood any beachfront property in Florida. Nevertheless, it's a volume of water equal to the submerged volume of the ship's hull up to the water line, and a buoyant force equal to that weight of water is still operating on the ship.

By the way, Archimedes didn't have an aircraft carrier at his disposal so, as the story goes, he used his own body. He filled his bathtub to the brim, climbed in, and realized that the weight of the overflow water on the floor had to be the same as his loss of weight—his buoyancy—in the water. History does not record the reaction of his landlady.

Nor did you ask this, but . . .

Exactly why does a hole in the hull of a ship make it sink?

Water rushes in through the hole because it is under a pressure, depending on how deep below the surface the hole is—the lower the hole, the harder the water rushes in. As the water enters the ship, it replaces an equal volume of air, thereby increasing the ship's weight and hence its overall density. When enough water has entered to make enough extra weight to overcome the buoyancy force, down she goes.

Fish Gotta Swim

While snorkeling around in the water, I saw a shell on the bottom that I wanted to collect. I tried to dive down, but it was devilishly hard to force my body down that deep. Fish were diving all around me. Why was it so easy for them to dive? Do they have something that I don't have?

The trouble is that you have something that they don't have: lungs.

In order to be at home suspended in seawater, in precisely neutral buoyancy without sinking or rising, a fish—or any other object—must have exactly the same overall density as the water. That is, it must weigh exactly the same as an equal volume of seawater (see p. 182). If it weighs more, it will sink to the bottom. If it weighs less, as most humans do, it will float to the surface and stay there. Ships, of course, are scrupulously designed to achieve the latter condition.

Bone and muscle are both denser than seawater, so just about any animal will sink unless it contains some very light stuff, such as an internal bag of gas, to compensate and reduce its overall density. We land creatures have lungs, and most fish have swim bladders—little gas-filled bags. Fish's swim bladders make up only about 5 percent of their entire volume, though, while our lungs fill most of our chest cavities. Lungs lower our overall density so much that our bodies are more buoyant than many kinds of wood.

Even if a fish happens to be denser than seawater, it can avoid sinking by constantly swimming. Similarly, when chasing seashells you could propel yourself downward by vigorously flicking your flippers, but alas! you're just not as proficient in the art of fin-propulsion as fish are. Even if you were, you would still have to work much harder, because you are burdened by those huge internal water-wings called lungs.

You didn't ask, but . . .

If a fish's density is just right for staying suspended in the water, how does it manage to rise or sink whenever it wants to change its altitude?

Of course, it can always swish its tail and swim to wherever it wants to go, but that's just a temporary solution. What it would really like to do is adapt its body to the pressure of the new depth, so that it can maintain its neutral buoyancy and rest there without constantly having to struggle up or down. It does that by adjusting its swim bladder.

When a fish travels to deeper water, the pressure on the fish increases because there is more water pressing down from above. This higher pressure compresses the swim bladder, making the fish denser than the exact value that it needs for neutral suspension. In order to remain effortlessly at that depth, it would have to reexpand its swim bladder. Conversely, when a fish ascends to shallower depths, it would have to compress its swim bladder in order to stay in neutral suspension without having to swim.

People used to think that fish did exactly that: expanded and contracted their bladders in order to adjust to different depths in the water. But scientists discovered that they just don't have the right muscles to do that. Surprisingly, what they do instead is change the amount of oxygen gas in the bladder. By adding or removing gas from the bladder, a fish can readjust its density to the exact density of water and remain effortlessly suspended without having to swim too much, no matter what the water pressure has done to its bladder size.

Where does a fish get the extra gas that it needs when it wants to stay at a greater depth? It takes oxygen out of its bloodstream and secretes it into the bladder. Where does it stash that extra gas when it wants to stay at a lesser depth? It absorbs some of the oxygen from the bladder back into the bloodstream. Ingenious!

Some poor fish don't have swim bladders. They are slightly denser than seawater and have to keep swimming to stay off the bottom. Mackerels and some species of tuna begin to sink as soon as they slow down. But the turbot just gives up and stays on the bottom.

So if you have to work hard in order to dive, take comfort in the fact that some fish have to work hard to keep from sinking.

Can Fish Get the Bends?

I have heard that fish can get the bends, just as scuba divers do when they stay down too long. Okay, I feel silly for asking, but how long can a fish stay under water without getting sick?

Fortunately, it isn't necessary to answer that question, because divers—and fish—don't get the bends (more accurately known as decompression sickness) from staying down too long. Divers get the bends by coming up too fast, but fish can indeed get the bends from other causes.

When the water pressure on a diver's body is reduced too fast, gas bubbles can form in his or her bloodstream. That hurts, to say the least. And yes, the same thing can happen to a fish, but not from swimming upwards too fast. It happens because of changes in the characteristics of the water itself.

Oxygen dissolves to some extent in water and in watery liquids such as blood and body tissue fluids. That's just great for the fish, of course, because they live on the oxygen that is dissolved in the water. But nitrogen, which is the major (78 percent) component of air and which is inert and useless to physiological processes, also dissolves in water and in blood. Ordinarily, this causes no problem for either fish or man, because we extract the oxygen that we need for metabolism and throw away the nitrogen via gill or lung. But if for some reason there is too much dissolved air in our bloodstreams, we may not be able to "undissolve" the excess nitrogen fast enough, and it can collect into actual bubbles of gas, blocking the circulation and destroying local tissues.

The amount of air that will dissolve in water at a certain temperature depends on the pressure: the higher the pressure, the more gas will dissolve (see p. 16). When a human diver goes down, the increased water pressure forces more oxygen and nitrogen into his or her bloodstream via the lungs.

The oxygen is no problem because the blood's hemoglobin eagerly latches onto it and delivers it to the cells. That's its job.

But when a diver surfaces and the pressure decreases, it would be very nice if the excess dissolved nitrogen could depart via the way it came in: through the lungs. Unfortunately, that can be a very slow process. Instead, when the pressure is reduced too quickly, the excess nitrogen gas just bubbles out of the blood, just as the carbon dioxide does when you release the pressure by opening a bottle of soda pop.

The answer for divers, of course, is to come up slowly and give the nitrogen a chance to leave the bloodstream gradually, molecule by molecule, exiting via the lungs.

If a fish were to swim rapidly upward from a great depth to the surface, the same thing might happen, except for two differences: One, fish have more sense than to do that; and two, something even more drastic would happen—the fish's swim bladder (see p. 186) would expand so much that it would crush the fish internally and kill it.

But we said that fish can get the nitrogen-bubble bends, and they can. Here's how.

Suppose that a fish is happily acclimated to its environment, swimming around in water that contains a certain amount of dissolved air. Its bloodstream will have adjusted itself to that same amount of nitrogen.

Now suppose that the fish wanders into water that for some reason (we'll get to the reasons) contains a lot more dissolved nitrogen than is normal at that temperature and pressure. Before long, its blood will also acquire that same abnormal amount of dissolved nitrogen. This is a precarious condition to be in, though, because at any moment some of that excess nitrogen can pop out as a bubble and the fish will get the bends. It can only relieve itself by heading for greater depths, where the added pressure will push the bubble back into the blood.

How might a fish find itself in water that contains an abnormal amount of nitrogen? It turns out that it needn't have anything to do with depth or pressure.

For example, a fish may be swimming in a river that contains the standard atmospheric-pressure's worth of dissolved nitrogen, when it happens upon a region of warmer water that has just been discharged by a factory or power plant. (Power plants inevitably throw away a lot of waste heat; see

p. 248.) By rights, the warmer water should contain less nitrogen, not more, because gases dissolve to a lesser extent in warm water than they do in cold (see p. 16). But if the plant's discharge water had never been given enough time to lose some of its extra nitrogen when it was heated—and remember that the evolution of nitrogen can be a slow process—then it will still be carrying more nitrogen than is normal for the river's normal conditions. The poor fish finds itself swimming in abnormally high-nitrogen water, and it gets the bends. That's one way in which power plants kill the fish in a river "merely" by discharging warm water into it.

Another example: Have you ever bought a couple of goldfish, taken them home, put them in a bowl of nice, fresh water and then watched them get sick and die? Well, here's what might have happened. Your tap water has lots of dissolved air in it because it is cold and it was probably even sprayed into the air at the waterworks to aerate it. Then you put it into the fishbowl, where it slowly warms up to room temperature. But it may still retain its cold-water load of nitrogen because, as noted above, evolving excess nitrogen can be a very slow process. The water then will still contain an abnormal load of nitrogen when you put the fish in it. Bends and death ensue.

Can anything be done about the power-plant fish-kills and the thousands of goldfish murders that are committed every day? Yes, and it's fairly simple. Just let the water stand for a long time before dumping it into the river or the fishbowl. Standing will permit any excess nitrogen to escape, and the water will come down to the just-right amount of nitrogen content for its temperature and pressure, and therefore the just-right amount for a fish at that temperature and pressure.

You didn't ask, but . . .

How do fish in deep ocean water get their oxygen? How much oxygen can there possibly be down there with the atmosphere so far above?

The oxygen doesn't come only from dissolved atmospheric air. You're forgetting about plants, which breathe in carbon dioxide and breathe out oxygen. Oceans contain an abundant variety of plant life, and the oxygen emitted by the plants dissolves directly into the water. By constantly swimming and passing large amounts of water over its gills, a fish can "vacuum up" a lot of oxygen, even if it isn't present in very concentrated amounts.

In areas where not enough plants exist to supply the fishes' breathing needs, they just take their business elsewhere.

Blowing Bubbles in the Air

Why are soap bubbles round?

Let's put it this way: You'd be pretty surprised if they were square, wouldn't you? That's because all of our experience since we were babies tells us that Mother Nature prefers smoothness. There just aren't many natural objects that have sharp points or jangling angles. The major exception is certain mineral crystals, which occur in beautifully sharp geometric shapes. That may be why some people believe that crystals and pyramids are endowed with supernatural powers.

But that is metaphysics, not science. Bubbles are round—spherical—because there is an attractive force called *surface tension* (see p. 7) that pulls molecules of water into the tightest possible groupings. And the tightest possible grouping that any collection of particles can achieve is to pack together into a sphere. Of all possible shapes—cubes, pyramids, irregular chunks—a sphere has the smallest amount of outside area.

As soon as you release a bubble from your bubble pipe or from one of those more modern gadgets, surface tension makes the thin film of soapy water assume the smallest surface area that it can. It becomes a sphere. If you hadn't deliberately trapped some air within it, the soapy water would continue to shrink down to a solid spherical droplet, as rain drops do.

But the air on the inside is pushing outward against the water film. All gases exert a pressure on their captors because they consist of freely flying molecules that are banging up against anything in their way (see p. 153). In a bubble, the inward surface-tension forces of the water film are exactly balanced by the outward-pushing pressure of the air inside. If they were any different, the bubble would either shrink or expand until they were equal.

Try to blow more air in to make a bigger bubble? That makes more air pressure inside. All that the water film can do to counterbalance the increased outward pressure is to expand its surface, making more inward-directed surface-tension forces. So it very cooperatively grows in size. But it must get thinner in the process, because there is only so much water to go around. If you keep blowing more air in, the film eventually won't have enough reserve water to spread out into a bigger surface, and the ultimate catastrophe occurs: Your bubble bursts.

Exactly the same thing happens with bubble gum, except that instead of surface tension as the inward, contracting force, it's the elasticity of the rubber in the gum. (Yes, rubber.) Elasticity, like surface tension, means "let's always try to assume the smallest possible shape."

You didn't ask, but . . .

Why do we need soapy water to blow bubbles? Why can't we use plain water?

In the strength of its inward-directed surface-tension force, water is the champion of all liquids. Its surface tension is so strong that water resists being stretched outward at all, even into the three-dimensional shape of smallest surface area: a sphere. Water knows that it can have an even smaller amount of surface area by simply lying flat and refusing to extend up into the third dimension at all. So pure water won't make bubbles of any shape—at least, not bubbles that will last for more than an instant.

Soap has the effect of reducing the surface tension of water (see p. 7). It weakens it enough so that the water's "skin" can be stretched into three dimensions.

Alcohol has such a low surface tension that it won't make bubbles at all. It would be like trying to blow bubbles with ordinary chewing gum that has virtually no elasticity.

Building a Better Wetter

Are all liquids wet?

No, all liquids are not wet. Even water is not always wet. It depends on who or what is the "wet-ee."

Make this inquiry of a linguist, however, and you'll be told that it is a foolish question. The word *wet* is so intimately related to the word *water* in the roots of our language that *wet* has always meant "coated with water." Water is "wet" by definition, and the opposite of *wet* is *dry*, which means "without water."

But language is a fickle facsimile of fact. The reason for the linguistic intimacy between water and wetness is simply that no other liquids were known by our primitive ancestors when they needed a word to describe the way you look when you come out of a river. After all, water is not only the most abundant liquid on Earth, it is the most abundant chemical compound of any kind. Even today, most people would be severely challenged to name two or three other liquids. Things like blood and milk don't count, of course, because their liquid parts are still water.

Innumerable other liquids do exist, however. In principle, any solid material can be melted into a liquid by heating it, and any gas can be condensed into a liquid by cooling it. It just happens that water exists in its liquid form over most of the temperature range at which life also exists. That's no coincidence, of course; life presumably began in the water, and liquid water is still essential to all forms of life.

Why, though, is this ubiquitous liquid wet? Why *does* it stick to us when we emerge from the river? Our primitive ancestors would have loved this explanation: It sticks to us because it *likes* us.

Putting it a little more scientifically, water molecules will adhere to those substances whose molecules hold some form of attraction to them. If there were no attraction between the molecules in a drop of water and the molecules at the surface of our skins, the water would just roll off. Our job, then, is to find out what those attractive forces may be.

At several other places in this book, we talk about the fact that water molecules are *polar*, and attract each other like tiny magnets (see p. 98). Water molecules are attracted to each other also by *hydrogen bonding* (see p. 98). If an alien substance comes along whose molecules are also polar or are also subject to hydrogen bonding, the water molecules will be attracted to them as if to their own. In other words, water will wet that substance.

Most proteins and carbohydrates, including the proteins in our skin and the cellulose in wood, paper, cotton, and other vegetable matter, are made of molecules with the right characteristics so that water molecules want to snuggle up to them. They will therefore be wetted by water. Other substances, however, such as oily or waxy materials, do not have either of the two necessary molecular characteristics to be wet by water.

TRY IT Dip a candle into a glass of water and you will see that water isn't necessarily wet. Water is sometimes "wet" and sometimes not, depending on what material we are tempting it to stick to.

What about other liquids? Are they always "wet"? We might wonder about such liquids as grain alcohol, isopropyl rubbing alcohol, gasoline, benzene, olive oil, and even liquid metals such as mercury. Like water, these liquids will wet materials to whose molecules they are attracted by a mutual attractive force. As far as human skin is concerned, the first five can find enough in common with "skin molecules" to adhere to them, and these liquids will wet you. But the atoms of metals have nothing in common with your "skin molecules" and won't wet them at all.

TRY IT If you ever get a chance to dip your finger into a pool of mercury, you will note that it comes out as dry as that candle that you dipped in water. (Don't linger over the mercury. Its vapor is toxic.) But dip a piece of clean copper or brass into the mercury and it will wet it eagerly, because metal atoms all have similar attractive forces and tend to stick together. If you have ever done any soldering, you know that the melted (metal) solder wets the metal parts that you are trying to join together.

NITPICKER'S CORNER:

Actually, wetness is a relative term. Some liquids are wetter than others; they will spread out and flow more readily over the surface that they are wetting.

Surprisingly, water isn't a very good wetter, as liquids go. Alcohol, for example is much wetter than water. That's because water's molecules adhere to each other so strongly that they tend to ignore other nearby molecules and won't adhere to them very readily even if they do have the right kind of molecular attractions.

TRY IT Sprinkle a few drops of water onto your umbrella and they will roll off, unless you force them to wet the material by rubbing them in with your finger. Sprinkle some alcohol onto the umbrella, though, and it will soak right in.

There are certain substances that, when added to water, will make it a better wetter. Soap is the most common one. (See p. 7 for how it works).

BAR BET All liquids aren't wet, and even water is sometimes dry.

The Hot-Freeze Paradox

Once and for all: Does hot water freeze faster than cold water? Some people swear that it does. Is there a definitive scientific answer?

Well, yes and no. Sorry about that.

This controversy has been raging ever since the early seventeenth century, when Sir Francis Bacon became a charter member of the Betcha-the-hot-water-freezes-first camp.

The only appropriate answer to this puzzle is, "It depends." It depends on precisely how the freezing is being carried out. Freezing water may sound like the simplest of happenings, but there are many factors that can affect the result. How hot do you call hot? How cold do you call cold? How much water are we talking about? What kind of container is it in? How much surface area does the water have? How is it being cooled? And exactly what do we mean by "freezes first"—a skin of ice on the surface or a solid block?

Let's listen to some of the yea-sayers and nay-sayers.

Naysayer: It's impossible! Water has to be cooled down to 32 degrees Fahrenheit (0 degrees Celsius) before it can freeze. Hot water simply has further to go, so it can't possibly win the race.

Yeasayer: Yes, but the rate at which heat is conducted away from an object is greater when the temperature difference between the object and the surroundings is greater. The hotter an object is, then, the faster it will cool off in degrees per minute. Therefore, heat will be leaving the hot water faster and it will be cooling faster.

Naysayer: Maybe. But who says the heat is leaving by being conducted away? There's also convection and radiation. See page 22 of the very book we're arguing in. Anyway, that would mean only that the hot water might catch up with the cold water in the race toward 32 degrees. But it could never pass it. Even if the hot water gets to be the same temperature as the cold water, they will thereafter continue to cool at the same rate. At best, they'll freeze at the same time.

Yeasayer: Oh, yeah?

Naysayer: Yeah!

Having reached the point of diminishing rationality, we may mediate the discussion by stating that thus far, the nays have it. Clearly, under absolutely identical, controlled conditions, hot water could never freeze faster than cold water. The problem is that hot water and cold water are inherently *not* operating under identical conditions. Even if we had two identical, open containers being cooled in exactly the same way, there are several factors that could possibly bring about a victory for hot water. Here are some of them:

• Hot water evaporates faster than cold water. If we start with exactly equal amounts of water (which is essential, of course), there will be less water remaining in the hot-water container when it gets down to rug-cuttin' time at 32 degrees. Less water, naturally, can freeze in less time.

If you think that evaporation can have only a trivial effect, consider this: At the typical temperatures of hot and cold household tap water (140 degrees and 75 degrees Fahrenheit), the hot water is evaporating almost seven times as fast as the cold. Over a period of an hour or two, a container of hot water can be substantially diminished by this rapid evaporation. Of course, as the hot water cools down, its rate of evaporation will gradually decrease. Nevertheless, on the way down it may well have lost a substantial amount of water.

• Water is a very unusual liquid in many ways. One of these ways is that it takes quite a lot of heat, relatively speaking, to raise its temperature by each degree. (Techspeak: Water has a high *heat capacity.*) Conversely, quite a lot of cooling is required to *lower* its temperature by each degree. If there is only slightly less water in a container, then, it may require substantially less cooling to get it down to freezing temperature. Thus, if the originally hot container has lost even a little bit of its water by evaporation, it may reach the freezing temperature quite a bit sooner than the water in the other container. That is, it could actually overtake the cold water and get to the finish line first. Moreover, once it is at the freezing point, water must have a great deal of extra heat removed from it to make it actually freeze into ice:

eighty calories for every gram (see p. 163). So again, a little less water can mean that a lot less cooling is necessary to freeze it.

- Evaporation is a cooling process (see p. 177). The faster-evaporating hot water will therefore be adding some extra evaporative cooling to whatever cooling process is operating on both containers. Faster cooling can mean faster freezing.
- Hot water contains less dissolved air than cold water does. Anything, including a gas, that is dissolved in water makes the water freeze at a lower temperature (see p. 92). The more air (or anything else) that is dissolved in water, the lower the temperature to which it must be cooled, in order to freeze. Having less air in it, the hot water doesn't have to be cooled to as low a temperature as the cold water does, and can freeze sooner.

However, this last, often-quoted argument doesn't hold water, so to speak. The lowering of the freezing point due to dissolved air amounts to only a couple of thousandths of a degree. Nevertheless (there's always a nevertheless), many people claim that when the water pipes freeze in an unheated house in the winter, it is usually the hot water pipes—containing previously heated water—that freeze first.

All things considered, then, it is quite possible that under some circumstances a bucket of hot water, left outside in the winter, will freeze faster than a bucket of cold. The claims of Canadians that they have seen it happen many times can be believed, even by scientists who "know better" and by other skeptics. The biggest and most probable effect in this case is the loss of water by evaporation. Extensive research, however, has not yet revealed why Canadians leave buckets of water outside in the winter.

But there are still a couple of hefty monkey wrenches in the works. First of all, a container of water doesn't cool uniformly throughout, until it all gets down to 32 degrees and suddenly freezes. It cools irregularly, depending on the shape and thickness of the container, what it is made of, the prevailing air currents, and several other variable factors. The first skin of ice to form at the surface of the water, therefore, may be a bit of a fluke, and may not be signaling that the rest of the water is ready to freeze. (The first ice to form will invariably be at the surface of the water; see p. 202.)

Second of all, believe it or not, water can be chilled well below 32 degrees Fahrenheit without freezing. It can be *supercooled*, yet it will not crystallize into ice unless some outside influence stimulates it to do so. The molecules may be all ready to snap into their rigid ice-crystal formation, but they need some final encouragement, perhaps in the form of a speck of dust that they can gather around, or maybe an irregularity on the wall of their container.

In view of these uncertainties, precisely when can we say that a given container of water has "frozen"? Our two buckets of water are racing without a clearly defined finish line.

All things considered, the best we can say is, "Hot water *can* freeze faster than cold water. Sometimes."

DON'T TRY IT If you are tempted to run right into the kitchen to fill two ice-cube trays, one with hot water and one with cold, and put them into your freezer to see which one freezes first, don't bother. There are just too many uncontrolled variables—things you can't control. You can get one result one time and the other result the next. That's the problem with people who say, "I know it works; I tried it," referring to anything from freezing water to curing warts. You just have to examine everything that could possibly affect the outcome. And there can be dozens of unsuspected angles, even to an apparently simple experiment like making ice cubes.

The Titanic Sank, but the Iceberg Kept Floating

Why do icebergs and ice cubes float? Aren't solids generally heavier than liquids?

Generally, yes. But water is an exception. As trivial as this question may sound, the answer is of life-and-death importance. If ice didn't float on water, we might not even be here to wonder why.

Let's see what would happen if ice sank in liquid water. In prehistoric times, whenever the weather got cold enough to freeze the surface of a lake, pond, or river, the ice would immediately have sunk to the bottom. Subsequent warm weather might not have melted it all because it would have been insulated by all the water above it. The next freeze would have deposited another layer of ice on the bottom, and so on.

Before long, much of the water on Earth, except for an equatorial band where it never freezes, would be frozen solid from the bottom up, and there might not be enough time in the warmer seasons to melt it all the way down. The primitive sea creatures from which we evolved might never have had a chance to develop. The world would be relatively barren of life.

The floating of solid water (ice) on liquid water is so familiar to us that we don't realize that it's really an unusual phenomenon.When most other liquids freeze, the solid form is *denser*, heavier, than the liquid, for a given volume. That's just what we would expect, because in solids the molecules are packed together more tightly than they are in free-flowing liquids, so naturally the solids will be heavier and will sink. Try it with a liquid that freezes at a conveniently comfortable temperature: paraffin wax.

TRY IT Add a piece of solid wax to some melted wax, and watch it sink. You'd get the same result with molten versus solid metals, oils, alcohols, and so on. But perform the scientific experiment of placing an ice cube in a glass of water, and you'll get the opposite result. The ice cube floats.

The reason for water's contrary behavior lies in the unique way in which water molecules are connected to each other in a piece of ice. They are connected by bridges (*hydrogen bonds*, see p. 98) between the water molecules. But consider what a bridge does. Brooklynites might say that the Brooklyn Bridge joins Brooklyn to Manhattan, but Manhattanites might insist that it *separates* Brooklyn from Manhattan. Well, in a sense they're both right, and that's literally what hydrogen bonds do to the water molecules in ice: They join the molecules together, but they also hold them a certain distance apart.

So instead of crowding together as tightly as the molecules in other solids do, the water molecules form a sort of open latticework. The molecules are now farther apart in the ice than they were in the liquid, so the ice takes up more space. A given weight of water occupies about 9 percent more space in the form of ice than it does in the form of liquid.

TRY IT Look carefully at the ice in your freezer's ice-cube tray. You'll notice that the cubes have little mountain peaks. In freezing, they had to expand, and being restricted on the sides and bottom, the only direction they could go was up.

If freezing water is confined so that it can't expand, it will burst the strongest container in the attempt. That's why a water pipe or an automobile engine can crack when the water inside freezes.

The bridges in ice don't form all at once at the instant of freezing. As we start cooling water down from room temperature, it gets denser and denser, just like any other liquid, because the molecules are slowing down and don't require as much elbow room. Most other liquids just keep getting denser and denser until they freeze, and the solid will be densest of all. But not dear old *aqua*.

Water becomes denser only down to a point. When it is cooled to 39.16 degrees Fahrenheit or 3.98 (let's call it 4) degrees Celsius, it starts going the other way—getting *less* dense as you cool it. That's because some of those bridges are beginning form. Finally, at 32 degrees Fahrenheit (0 degrees Celsius), all the rest of the bridges snap into place, the water freezes into ice, and the density falls suddenly to a value that's lowest of all. That's why ice will float on water of any temperature.

The fact that water has a maximum density at about 39 degrees Fahrenheit has further significant consequences for living things. When cold weather cools the surface of a freshwater lake, the water at the surface becomes denser and sinks. New water takes its place, gets cooled and sinks. This goes on until all the water in the lake has had a chance to be cooled to its maximum possible density—at 39 degrees—and to sink. Only then can the surface water be cooled down those last seven degrees to form ice at 32 degrees Fahrenheit.

By the time a surface layer of ice can form on a lake, then, all the water in the lake will already be at a temperature of 39 degrees. No matter how cold the weather becomes, any water that gets colder than 39 degrees stays at the top (because it's lighter), and the fish below can never get any colder and freeze. That's another reason that water's peculiarities are responsible for permitting the existence of life on earth.

NITPICKER'S CORNER:

In real-life bodies of fresh water, temperature fluctuations, winds, water currents, and other mixing phenomena will mess up these tidy arguments about neat layers of water temperatures. But *all other things being equal* (the universal cop-out), the principles we've outlined above will prevail.

In the oceans, however, it's a slightly different ball game. Because of all the salt it contains, seawater doesn't have a maximum density at 39 degrees. As its temperature goes down, it just keeps getting denser and keeps on sinking, all the way down to its freezing temperature. In order for ice to form on the ocean surface, *all* the water first has to get down to the freezing point. That happens only during a long, hard winter near the North or South Pole.

BAR BET Show me a freshwater pond with a layer of ice on the surface and I'll tell you the temperature of the water at the bottom, without a thermometer.

On the Level

How does water "seek its own level"? I mean, how does one part of the water know where the levels of all the other parts are, no matter how distant?

It requires no psychic powers; just gravity.

"Water seeks its own level" is a catch phrase that was probably uttered by a Greek philosopher two thousand years ago, and people have been parroting it ever since. In plain language, it means that water will lie flat whenever it can.

If a body of water—anywhere from a bucket to a bathtub to an ocean—is left undisturbed, it will quickly settle down into a perfectly flat surface, no matter how wavy it started out. It will find the mathematically exact compromise level, averaging out the highs and lows as accurately as any army of surveyors with transits could do. But how indeed does a "hill" know that it must fall, and a "valley" know that it must rise?

It all happens because water (like other liquids) is *incompressible*. You can't force a liquid to occupy a smaller amount of space by pushing on it, the way you can with a gas. The reason is that the molecules of liquids are already as close together as they can get, and no amount of pressure (within reason) will make them crowd together any closer.

What goes for pushes also goes for pulls. Suppose there's a "hill" on the water's surface. Gravity is trying to pull it down, but its molecules can't oblige by packing down any more compactly; all they can do is spread out sideways into the lower-altitude surrounding territory. As a result the hill has disappeared and a valley has been filled.

Of course, gravity is also pulling down on the valley water, but it is already as low as it can go. In order to dig itself any deeper it would have to send the excavated water somewhere else: uphill. And that, of course, would be contrary to the force of gravity.

A hill of earth rather than of water would behave in the same way if its molecules could flow past one another as easily as water's molecules can. A hill of sand is an intermediate case; its grains can flow to some extent, so a hill of sand that is too high will "seek its own level" just as water does, although it may never achieve what it seeks. Water is less like a hill of sand than a hill of marbles.

Okay, so you knew all that. But here's a truly startling application of the same principles: the *sight glass*. You've seen them. On the outside of a boiler or other opaque container of water there is a vertical glass tube, which is connected to the water inside. You can't see the water level inside the boiler, but you can tell how high it is because it is at exactly the same level as the water in the external glass tube. How does the water in the glass tube know where the water level is inside the boiler?

Well, if the water inside the boiler were temporarily higher than the water in the sight glass, it would level itself down, just as the "hill" did a few paragraphs back. In this case, though, the excess hill-water has no valleys to flow into; it has no place to go except into the glass tube. Result? The tube level goes up and the boiler level goes down. The flow will stop when they reach the same level. Same thing the other way around: if the level in the glass tube were temporarily higher than the level inside the boiler. Either way, they come to exactly the same level.

TRY IT Does your kitchen have one of those plastic-cup gravy separators that look like miniature watering cans? The kind that lets you pour off the juices from the bottom, leaving the top layer of fat behind? It's a great substitute model for a boiler and sight glass. Put some water in it and notice that no matter what the water level is inside the cup (the "boiler") and no matter how you tilt it, the water finds exactly the same level in the transparent spout (the "sight glass").

Really Haute Cuisine

Why does a boiled egg cook faster in New York City than it does in Mexico City?

It would be great fun if we could attribute the difference to the Big Apple's hustle or to Mexico's mañana attitude. Unfortunately, we can't. The difference doesn't even have anything to do with the eggs. It's the water. But it's not what you think.

When it's boiling, water in New York is a little hotter than water in Mexico City. And hotter water will get an egg to a given state of doneness in a shorter time.

A little thought will show that the biggest difference between New York and Mexico City, apart from the relative difficulty of finding a good corned beef sandwich, is the altitude. The average stove in Mexico City is 7,347 feet higher than the average stove in New York City. And the higher the altitude, the lower the temperature at which water boils.

How much lower? If pure water boils at 212 degrees Fahrenheit (100 degrees Celsius) in New York City (and it may not; see p. 210), it will boil at only 199 degrees Fahrenheit (93 degrees Celsius) in Mexico City. Not a huge difference, but your exemplary three-minute New York City egg will certainly take longer to create in Mexico City.

The reason is simple, once you realize what boiling consists of: water molecules becoming energetic enough to break away from their brethren in the pot, then gathering together into rising bubbles, and finally flying off into the air as steam (see p. 45).

In order to escape, water molecules have to have enough energy—that is, they have to be at a high enough temperature—to overcome two separate forces: (a) They have to break apart the stickiness that holds them together in the liquid, and (b) they have to overcome the pressure that the atmosphere is applying to the surface of the water. That pressure is caused by air molecules that are continually bombarding the surface of the water like a barrage of ricocheting hailstones.

The sum-total force of those collisions is transmitted through the water to every molecule within it. Molecules at the surface can just fly off into the huge spaces between the air molecules, but those in the interior of the water must overcome this sum-total pressure in order to get out.

The stickiness of liquid water molecules to each other is the same, of course, whether they're part of a Manhattan or a margarita. But atmospheric pressure is another story. In Mexico City, the air is only 76 percent as dense as it is at sea level. That means that only about three-quarters as many air molecules are bombarding the surface of the water every second. The water molecules are therefore able to muscle their way upward and boil off without having to have quite so much energy: that is, without having to get quite so hot.

An extreme: The highest point on this planet is Mount Everest, which is 29,028 feet (8,848 meters) above sea level. At this altitude, the atmospheric pressure is only 31 percent of what it is at sea level, and the boiling temperature of water is only 158 degrees Fahrenheit (70 degrees Celsius). That's not hot enough to cook much of anything, no matter how hungry you may have gotten while climbing.

You didn't ask, but . . .

Does that mean that we could make water boil hotter if we artificially increased the pressure on it?

Absolutely. That's precisely what a pressure cooker does. Let's clamp a tight-fitting lid with only a small hole for escaping steam onto a cooking pot. Then we'll place a weight on top of that hole to keep a certain, calculated amount of the steam pressure in, instead of letting it escape freely into the atmosphere. Or else we can use some sort of pressure regulator to fix the pressure at a predetermined value. The pressure of the "atmosphere" inside the pot will then be maintained at that higher value.

At a typical pressure-cooker pressure of ten pounds per square inch (0.70 kilogram per square centimeter) above normal atmospheric pressure, the boiling temperature—and hence the temperature of the steam inside—is 240 degrees Fahrenheit (115 degrees Celsius). That's hot enough to make short work of any otherwise long-simmering dish, such as a stew. Moreover, the space inside a pressure cooker is filled with steam, which is a much better conductor of heat than air (see p. 26). Thus, any heat anywhere in the pot will be conducted into the food more efficiently than if the pot were filled with air. This also makes for faster cooking.

Teapot in a Tempest

If the boiling temperature of water depends on altitude because the atmospheric pressure differs, won't it also depend on the weather? According to weather reports, the atmospheric pressure is always changing, even in a single location.

Right you are. But the weather has only a small effect on the boiling temperature of water.

When people go around saying that water boils at 212 degrees Fahrenheit (100 degrees Celsius) at sea level, they're speaking rather loosely. The standard definition of the boiling temperature of pure water says nothing about sea level. It is defined in terms of a specific atmospheric pressure: 29.92 inches (760 millimeters) of mercury (see p. 153), which is a typical, but hardly guaranteed, value for sea-level locations. Every TV weather fan knows that the air pressure changes as the weather changes, whether you live by the seaside or anywhere else. So the temperature of boiling water will indeed depend on whatever the weather conditions happen to be at the time.

Quite arbitrarily, scientists have chosen exactly 760 millimeters of mercury as the standard pressure they call one *atmosphere*. (That strange-looking number, 29.92 inches of mercury, is simply a quirk of conversion from millimeters to inches: 25.4 millimeters per inch.) The boiling temperature at that standard pressure is called the *normal* boiling temperature or the normal boiling point. *That's* what 212 degrees Fahrenheit or 100 degrees Celsius really is.

While knowing these facts might impress your friends, the effect of atmospheric pressure on the boiling temperature of water isn't big enough to worry about. Even if you were brewing a cup of tea while sitting smack in the eye of a hurricane, where the pressure might drop as low as 28 inches or 710 millimeters of mercury (the world's record low is 25.9 inches or 658 millimeters), the boiling temperature would only go down to 208 degrees Fahrenheit (98 degrees Celsius). It's comforting to know that your tea would still be hot enough.

Skating on Thin . . . Water

The human record for running speed is about twenty-three miles per hour. But for ice skating it's more than thirty-one miles per hour. Obviously, sliding on the ice must increase one's speed. But why is ice so great for sliding? What makes it so slippery?

Actually, solid ice itself isn't slippery. There's a thin film of liquid water on its surface that the skaters are sliding on.

Solids in general aren't slippery because their surface molecules are tied tightly together and can't roll around like ball bearings. The molecules of liquids, on the other hand, are free to move around, so liquids are generally slipperier than solids (see p. 101). A little water on a tile or concrete floor can turn it into an accident-lawyer's dream.

But what scientists can't quite agree upon is exactly what creates the liquid film on the surface of ice. Obviously, it must come from slight melting, but what makes the ice melt?

Two explanations—pressure melting and friction melting—have been slugging it out during the more-than-a-hundred years that people have been trying to explain this simple everyday phenomenon.

The pressure-melting camp maintains that it's the pressure of the ice-skate blade on the ice (or the ski on the snow, for that matter) that does the melting. There's no doubt that ice will melt if you apply pressure to it because solid ice occupies a bigger volume than liquid water does (see p. 202), and if you press hard enough on a piece of ice, you can force it to collapse down into its smaller-volume form: liquid water. The weight of an ice skater, concentrated on a tiny area the size of a skate blade, can amount to a pressure of thousands of pounds per square inch. The problem, though, is that even this intense pressure isn't enough to do the necessary amount of high-speed melting, especially when the ice is quite cold, because that's when its molecules are most tightly set in their rigid ways.

But wait a minute. Rubbing any two solids together, even a skate blade and a piece of ice, is bound to create friction, and friction creates heat. According to the friction-melting camp, this frictional heat is enough to melt a continuous liquid streak as the skate blade or ski skims along the ice or snow.

The best evidence today seems to favor friction melting, assisted by some pressure melting at temperatures that are not too far below the freezing point.

TRY IT Using a towel to avoid melting it, pick up an ice cube from your freezer compartment or take out a whole tray of ice. Gently feel the ice by passing a finger across its surface. Don't rub too hard. You'll find that the ice isn't slippery at all until your body heat and friction heat have had a chance to warm up the surface and melt it a little.

BAR BET Clean ice is not slippery.

(Don't actually try this test in a bar, however. The bartender's ice probably isn't cold enough; it will be wet and slippery from the outset.)

The Hose Knows

It must be some ancestral memory from cave-man days, but we all seem to know intuitively that water will put out a fire, and we never question it. Well, why does water put out a fire?

Before we get any further, note well: Water must never be used on an electrical fire or on an oil or grease fire. Reasons: Water conducts electricity and can lead it elsewhere, perhaps to your very own feet. And because water won't mix with oil or grease (see p. 98), it just scrambles it around and spreads the fire.

Fire needs three things to survive: fuel, oxygen, and—at least initially—a temperature high enough to ignite the fuel and get the combustion reaction to begin. After that, the reaction gives off more than enough heat to keep things going.

Obviously, the first thing to do would be to remove the fuel. Nothing to burn, no fire. But water can't do that, so it attacks the other two essentials: the oxygen and the temperature.

A deluge of water from a bucket or hose can smother the fire as if it were a blanket, simply by blocking out the air. A thin layer of water, even for a short time, can do the trick. No air, no oxygen, no fire.

Water can also lower the temperature of the material

that's burning. Every combustible material has a minimum temperature that it has to reach before it will ignite and burn. If the water cools the material below that temperature, voilà! no more burning. Even hot water is well below the temperature at which most things can burn.

A deluge isn't necessary. Water from a sprinkler can put out a fire, even though it leaves a lot of oxygen available between the rain drops. So it must work by lowering the temperature. Remember how cooling it is to run through the lawn sprinkler?

A sprinkler lowers the temperature in two ways. First, water in fine droplets tends to evaporate quickly, and evaporation is a cooling process (see p. 177). Second, water has a peculiarity that makes it much better than any other liquid for dousing fires: It is a heat-sucking glutton. Water has a huge appetite for heat. A pound of water absorbs 252 calories of heat before its temperature goes up by a single degree Fahrenheit.

Is that a lot? Well, contrast it with the amount of heat it takes to raise the temperature of a pound of several other substances by one degree Fahrenheit: Mercury requires only 8.3 calories; benzene, 63 calories; granite rock, 48 calories; wood, 106 calories; and olive oil, 118 calories.

The moral is that a little bit of water can take away a lot of the fire's heat before boiling away and leaving the premises as steam. Water is therefore an extremely effective cooling agent. That's why it is used in automobile cooling systems. Of course, being cheap doesn't hurt.

You didn't ask, but . . .

Why won't wet things burn?

As we were saying, water is a champion heat absorber, without getting very hot in the process. When you put a flame to something that's wet, the water soaks up the heat like a sponge, preventing the object itself from ever getting hot enough to ignite.

TRY IT This one will astound you. Put a little water in an unwaxed paper cup (not a foam cup) and figure out a way to prop it up so that you can set a candle under it. (Maybe you can sit it on a wire rack that's bridging two coffee cans.) Place a lighted candle under the bottom of the cup. The cup won't burn, but after a while the water will get hot enough to boil, having absorbed the heat from the paper as fast as the candle gave it off. Even when the water boils, it will never get hotter than 212 degrees Fahrenheit or 100 degrees Celsius (see p. 45), which isn't anywhere near hot enough to ignite the paper. The candle's heat goes into boiling the water, not into heating up the paper.

Why Bars Are So Noisy

Why do ice cubes snap, crackle, and pop when I put them in my drink?

If you listen with a linguist's ear, you'll find that the ice isn't actually popping, which implies a certain hollowness. But it certainly does snap and, on occasion, crackle.

First, the snap. When you plunge a cold ice cube into a warmer liquid, the water warms up parts of the ice cube, which tends to make those parts expand slightly. This places a stress on the ice crystal, because ice has a very rigid structure and it can't just expand here and there at random. The only way the crystal can relieve these stresses is to crack. That's the snap you hear.

Next, the crackling, which sounds like a rapid series of tiny explosions. And that's exactly what it is. Unless you've made your ice cubes out of boiled water (see below), there was some dissolved air in the water that went into your ice tray—or, if you're one of those most fortunate of human beings, into your automatic ice maker. As the water froze, there was no room for the air in the rigid, solid structure that is ice, so the air had to settle out into tiny, isolated bubbles. These bubbles are what makes the ice cloudy instead of crystal clear.

Now put that bubble-filled ice cube into a drink. The water works away at melting the surface of the cube, eating its way in deeper and deeper. As it goes, it encounters an air bubble. When that bubble was formed, it was freezer-cold air. But now it is being warmed up by the advancing water, and it wants to expand. It can't expand, however, until its imprisoning ice wall has been thinned enough to allow it to break through. When it does, Crack! It explodes its way out. Thousands of those tiny breakouts, happening all over the ice surfaces, make a faint crackling or sizzling noise.

The crackles of icebergs and glaciers as they move south into warmer water can be heard loud and clear by the inhabitants of Arctic-prowling submarines.

TRY IT Boil some water for several minutes to get most of the dissolved air out of it. Let it cool, pour it into an ice-cube tray, and freeze it. You'll find that there won't be many bubbles in the ice cubes. (Compare one with an ordinary ice cube by holding them up to a strong light.) When you put the boiled cubes into a drink they may snap, but they won't crackle much. You'll enjoy a relatively quiet drink.

7

. . . And That's the Way It Is

We have looked closely at well over a hundred everyday happenings and have seen what makes them happen. But is that what science is? Questioning each individual happening and finding a unique explanation for it, only to move on to the next, and the next? Not at all.

There are certain general principles that underlie and relate many of the situations that we have discussed. The extensive cross-referencing shows how interrelated many of our questions really are. It would have been more logical and much more efficient if I had first explained the general principles and then shown how they applied to the various aspects of your everyday life. But then we wouldn't have had a question-and-explanation book; we would have had a textbook. And that's not what you wanted.

Nevertheless, those general principles do exist; scientists call them *theories*. When a theory has been thoroughly tested and has passed with flying colors, it may achieve the exalted status of a *Law of Nature*. A Law of Nature is simply an elegant way of saying, "This is the way the world works. We may not know *why* it works that way, but that's the way things are, like it or not."

You've heard of Newton's Law of Gravitation and perhaps

of his Laws of Motion. But you may not have heard of the three Laws of Thermodynamics, the powerful laws that govern changes in energy. And nothing happens—*nothing*—without a change in energy.

Science has come up with many other general descriptions of the way things are. The last section of this book invokes these general principles to answer some fundamental questions about energy, gravity, mass, magnetism, and radiation; from seeing in the dark to seeing through lead. Along the way I'll sneak in a little propaganda for the metric system.

This section—and the book—ends with the seemingly childish, but most profound question of all: "What makes things happen or not happen?" The Second Law of Thermodynamics will give us the answer.

More Heat Than Light

I don't understand infrared radiation. How can it be used to see in the dark? People sometimes call it "light" and sometimes call it "heat." Which is it?

Strictly speaking, it's neither. It's not light because we can't see it, and it's not heat because it contains no substance that is capable of being hot. I like to call it "heat in transit." We'll see why.

Infrared radiation is nothing more than a certain segment of the broad variety of *electromagnetic radiations* that are being showered down upon us by the sun. Electromagnetic radiations are waves of energy, traveling through space at the speed of light. Being pure energy, they are distinguished from those so-called radiations that are actually streams of tiny particles, such as some of the radiations that are emitted by radioactive materials.

Electromagnetic radiations differ from one another only by their energies. The lowest-energy radiations are radio waves and the highest-energy ones are called gamma rays. In between we find (going up in energy) microwaves, infrared radiation, visible light, ultraviolet rays, and X rays. Gamma rays come mostly from radioactive materials. Radio waves, microwaves, and X rays, we have to make ourselves. The rest of this *spectrum*—this spread of electromagnetic radiation energies—is provided in abundance by Old Sol.

In order to observe electromagnetic radiations, we must have just the right sort of instrument, attuned to the exact radiation energy we want to detect. For a certain very small part of the sun's spectrum, we have a marvelous instrument called the human eye. Not surprisingly, the part of the spectrum to which this instrument is sensitive is called visible light. For radio waves and microwaves, we need an antenna to collect them and electronic circuits to convert them into something that we can see or hear. For X rays and gamma rays, we need instruments like Geiger counters and other paraphernalia that nuclear physicists use.

Infrared radiation (*infra-red* means "below the red" in energy) is just outside the energy region that human eyes can detect. That's why we can't properly call it light. We must detect infrared radiation by its effect on things. And its major effect is its ability to heat things up.

Radiations of various energies have different kinds of effects when they hit matter, when they strike the surface of any substance. In general, there are three possibilities: The radiation can bounce off, it can be absorbed, or it can go straight through.

Visible light bounces off most substances, while X rays generally go straight through. But infrared radiation has just the right amount of energy to be absorbed by the molecules of a wide variety of substances. When a molecule absorbs energy, it becomes, of course, more energetic. It jiggles, rotates, flaps its atoms, and tumbles around more than it did before. And an energetic molecule is a hot molecule (see p. 236).

So when infrared radiation shines on something, it makes that thing warmer. The radiation itself isn't "heat" until it reaches some kind of substance and is absorbed. That's why I call it "heat in transit."

You're most likely to see infrared radiation being put to use in two common applications: heat lamps and infrared photography.

Heat lamps are used in restaurants in an attempt to keep your food warm from the time the dish is assembled in the kitchen until your server returns from what appears to be an extended vacation. The lamps are designed to put out most of their "light" in the infrared region of the spectrum, although some of it spills over into visible red light.

Infrared photography—photography "in the dark," meaning without *visible* light—is based on the fact that as warm objects lose their heat, they emit some of it in the form of infrared radiation (see p. 22). This radiation can be detected by special photographic films or by phosphorescent screens. The warm objects are thus rendered visible.

People themselves are warm, infrared-emitting objects, and are often the targets of those see-in-the-dark snooper devices that work on this principle.

Don't Throw Away That Lead-lined Blouse, Lois

Why can't Superman see through lead with his X-ray vision?

He could if he really tried. It's just that his inventors, Jerry Siegel and Joe Shuster, told him that he can see through anything but lead, and like any good cartoon character, he faithfully obeys his creators.

Siegel and Shuster's idea seems to have been that X rays can't penetrate lead. Otherwise (the reasoning goes), why do X-ray technicians hide behind a lead-lined wall when they zap you? Why do dentists drape you with a lead apron when they shoot your teeth?

Lead is indeed used as radiation shielding throughout the world of nuclear research and technology. But the truth is that there's nothing special about lead at all. It simply does the job more cheaply than other materials.

X rays are just one kind of electromagnetic radiation, pure energy that zips through space at the speed of light. Other kinds of electromagnetic radiation that are more familiar outside the doctor's office are light itself, the microwaves that cook your food, and the radio waves that carry all those programs to our radios and televisions.

All of these energy waves are vibrating up and down and sideways as they fly along. In fact, their energy *consists of* these vibrations: a higher frequency of vibration—more vibrations per second—means a higher radiation energy.

They line up this way in order of increasing energy: AM radio, short-wave radio, television and FM radio, radar, microwaves, light (both visible and invisible to humans), X rays, and gamma rays, the last of which are emitted by radioactive materials.

Being of such high energy, you might expect (as if you didn't know) that X rays are very penetrating radiations. They go through flesh like a bullet through Jell-O. Bones block them just enough to throw diagnostic shadows on a sheet of photographic film. The bad news, though, is that X rays, like gamma rays, are *ionizing* radiations. That is, as they plow through atoms of flesh, bone, or anything else, they knock out electrons, leaving behind ions—atoms that are missing some of their electrons.

And without going into detail, let it be said that atoms that are playing without a full deck of electrons are—to mix a metaphor—loose cannons in the chemical game of life. They can disrupt our body chemistry in strange and unhealthful ways. That's why we want to shield ourselves from X rays and other ionizing radiations, such as those that come from radioactivity.

What, then, shall we use to stop X rays? Anything that offers lots of atoms with lots of electrons to knock out, because each time a beam of X rays dislodges an electron from an atom, it costs it some of its energy. So the more atoms with lots of electrons we can put in their way, the sooner the rays will lose all of their energy and stop. The best X-ray stopper, then, is whatever substance has the largest number of electrons per atom and is the most densely packed—with the largest number of atoms in each cubic inch.

Uranium would be just dandy. It has ninety-two electrons per atom and is nineteen times as dense as water. Gold would be great, too: seventy-nine electrons per atom and slightly denser than uranium. And then there's platinum: seventy-eight electrons per atom and twenty-one times denser than water. But alas! These substances are all too expensive. And who wants to escape X rays by hiding behind a wall of radioactive uranium, anyway?

So it all comes down to how many electrons per cubic inch you can buy for a buck. Lead fits the bill better than any other material. It has eighty-two electrons per atom, is 11.35 times denser than water, and a dollar will get you about ten pounds of it.

(In case you were wondering, there are 4×10^{25} electrons in a cubic inch of lead. That's a four, followed by twenty-five zeroes.)

But some X rays will always get through a sheet of lead or anything else, no matter how thick. It's just that the thicker the layer, the fewer will get through. In theory, a beam of X rays can never be stopped completely by *any* thickness of *any* material. We can only reduce the beam to a relatively harmless level.

Of course, you can use an even cheaper, if less effective, X-ray stopper than lead; you'll just need more of it. A thick concrete wall, for example, will do the same job as a relatively thin sheet of lead, even though concrete isn't anywhere near as good an X-ray absorber, thickness for thickness. If you have lots of room for shielding, you can even use the cheapest of all materials: water. Only ten electrons per molecule, but with enough of it between you and the X-ray source, you're safe.

Siegel and Shuster may have known all of this, but admitting it would have spoiled a great literary gimmick. So Lois Lane can rest easily after all in her belief that lead-lined clothing foils X-ray vision.

Until old mild-mannered Clark wises up.

Cool as a . . . Rutabaga

Why are cucumbers "cool as a cucumber"? I've read in both a cookbook and a food magazine that cucumbers are always twenty degrees cooler than their surroundings. What makes that happen?

Twenty degrees, eh? Well, let's just see about that. (I'll assume that we're dealing with Fahrenheit cucumbers, rather than Celsius.)

If cucumbers are always twenty degrees cooler than their surroundings, let's put a cucumber into a barrel with a whole bunch of other cucumbers and wait to see what happens. Will they fight it out, each one trying to be twenty degrees cooler than its neighbors? Have you ever seen a bushel of cucumbers suddenly freeze itself solid for no apparent reason?

Or, how about this: If cucumbers are always twenty degrees cooler than their surroundings, let's build a big box out of cucumbers and keep all of our wine nice and cool in it, at maybe fifty-five degrees Fahrenheit. And why stop there? Let's build a smaller cucumber box and put it inside the first one, thereby lowering the temperature by another twenty degrees and we'll keep our beer in it at a nice thirty-five degrees. And no ice required, thank you, because with one more box we can get down to a temperature well below freezing and make our own. With enough boxes-within-boxes, we could build a refrigerator that would freeze Hell itself. And all without even having to plug it in.

We have just violated the most basic law of physics: the First Law of Thermodynamics, more familiarly known as the Law of Conservation of Energy. For here we have a substance—cucumber flesh—that must be constantly shooting off heat energy into its surroundings. That's the only way an object would be able to stay cool: by constantly throwing off any heat that might flow naturally into it from nearby objects. Since heat is energy, the cucumber flesh is, in effect, an inexhaustible fountain of energy. Free of charge. No need to burn coal or oil or to put up with the problems of nuclear energy. Why, we can use cucumber energy to generate electricity, to propel smog-free automobiles, to irrigate the deserts to grow more and more cucumbers! Why, we can . . .

The only thing we can't do is stop people from putting silly things in books. And of course, the totally fabricated number of twenty degrees is irrelevant. Automatically cold cucumbers—or automatically cold anything else—simply can't exist. Nothing can permanently maintain even a slightly different temperature—colder or hotter—from its surroundings, unless we supply or remove energy to or from somewhere else. That's why we have to plug in our kitchen appliances; we use electrical energy from the local power plant to pump heat energy out of our refrigerators and to pump heat energy into our ovens.

But you say you picked up a cucumber that hasn't been in the refrigerator and placed it against your forehead and it really did feel cool. It most certainly did. But that's because the cuke is cooler than your ninety-eight-degree skin, not because it's cooler than the seventy-degree room.

TRY IT Leave an unrefrigerated cucumber and a potato in the same location for several hours. Cut them and hold the cut surfaces against your forehead. They'll feel equally cool. Prove that they're really the same temperature by thrusting a meat thermometer into each.

Except for variations due to such things as air currents and sunshine streaming through a window, every object in a room is the same temperature. Unless you've turned the thermostat up to ninety-eight-point-six degrees, they'll all feel cool compared with your skin.

When any two objects are in contact, heat will flow spontaneously from the warmer one to the cooler one. So when the cuke—or any other room-temperature object—sucks out some of your hard-earned forehead heat, you feel the loss as a cool sensation.

Scientifically speaking, of course, there's no such thing as coldness; there are only various degrees of heat. The words "cool" and "cold" are mere linguistic conveniences. And so is the expression "cool as a cucumber." It's so much more fun to say than "cool as a rutabaga."

BAR BET A cucumber isn't one bit cooler than a potato.

What Was Einstein Really Getting At?

I know that Einstein's equation $E=mc^2$ is terribly important, and that it has something to do with the atomic bomb. But what does it really mean to us folks on the street?

Frankly, not a hell of a lot. But that's not to say that it isn't one of the most momentous realizations ever to dawn upon the human mind. Although it has to do with things that are happening right under our noses every day, they are much too small to notice except when brought to our attention by that bomb that you mention, which is assuredly one of the most effective attention-getting devices of all time.

This most famous of all equations was first put on paper by Albert Einstein in 1905 as one small part of his theories of relativity. Among many other things, Einstein discovered that there is an intimate relationship between mass and energy. (Energy is the ability to make things happen, while mass is essentially the weight of a material object.)

Intuitively, we would love to believe that energy is energy and objects are objects, period. But Einstein discovered that energy and mass are really two different but interchangeable aspects of the same universal stuff, which for want of a better term we may call *mass-energy*. Einstein's astoundingly simple little equation is the formula for determining how much energy is equivalent to how much mass, and vice versa.

(For the mathematically unchallenged: If m stands for an amount of mass and E stands for the equivalent amount of energy, the equation says that you can determine that amount of energy simply by multiplying m by a number represented as c^2. The number c^2 is incomprehensibly huge—it is the square of the velocity of light—so you can get an enormous amount of energy from a minute amount of mass.)

The reason that Einstein's equation isn't very relevant in everyday life (with one major exception that we'll mention) is that all of our common, daily energy-producing activities, such as metabolizing our food and burning coal and gasoline, are purely *chemical* processes, and in all chemical processes, the amounts of mass that the energy came from are utterly minuscule.

How minuscule? Well, even if we explode a pound of TNT, which you will agree is a chemical process that releases a pretty fair amount of energy, all of that energy comes from the conversion of only half a billionth of a gram (twenty trillionths of an ounce) of mass. If we could weigh the TNT before the explosion and then gather up all the smoke and gases after the explosion and weigh them, we would find that they weigh half a billionth of a gram less.

That is far, far beneath our notice. We can barely measure such a tiny difference in weight with the world's most sensitive scales. So while Einstein's equation applies without exception to all processes that involve energy—and don't let anybody tell you it doesn't—it is of no consequence whatsoever in our everyday lives.

That goes for all *chemical* processes. *Nuclear* processes, on the other hand, such as the nuclear fusion reactions that go on in the sun and the nuclear fission reaction that goes on in an atomic bomb, are quite another story. Because virtually all the mass in the world resides in the incredibly massive nuclei of atoms, much greater amounts of energy can be released, atom for atom, in a nuclear process than in a chemical process. Billions of times greater (see p. 228).

What really makes an atomic bomb the champion of all earthly energy releasers, however, is something called a *chain reaction*. That's a process in which each atom's worth of reaction makes two more reactions, and each of those two makes two more, and each of those four makes two more, and each of those eight makes two more and so on, until we have an incredibly huge number of atoms undergoing reaction, all sparked by a single-atom "starter" reaction. When you have an incredibly huge number of atoms reacting within an incredibly short period of time, each one giving off as much energy as a billion ordinary chemical reactions, you've got yourself one hell of an explosion.

Chain reactions aren't all bad. If we control the speed at which a nuclear fission chain reaction multiplies itself, we have a nuclear reactor. In a nuclear reactor, the energy is given off gradually enough to generate heat to boil water to make steam to drive turbines to drive generators to make electricity to light the lamp you may use to read this book.

That's what's in it for us folks in the street.

BAR BET Mass is converted into energy during ordinary chemical reactions.

Only well-schooled people will bite on this one; you might even hook a chemistry teacher. Chemists are so used to ignoring the tiny mass changes associated with chemical reactions that they believe there simply *are* no mass changes, and that's what ends up being taught in the schools. Clinch your argument by reminding your adversary that Einstein never said, "$E=mc^2$, except in chemistry class."

How Fat Atoms Lose Weight

I can understand that coal and oil must contain energy, because the energy comes out as heat when we burn them. But how do we get energy out of uranium? Does it burn?

If by "burning" you mean combustion—a chemical reaction with the oxygen in the air—no. But if you mean do the uranium atoms get used up, yes.

You're right about coal, oil, and uranium containing energy. Actually, every substance contains a certain amount of energy. It is inherent in the unique arrangement of its atoms and how they are held together. If the atoms are held together very tightly, they are in a relatively satisfied state and have low energy. If they are held together only loosely, they have more potential for change; they contain more potential energy.

The atoms in nitroglycerin, for example, are very loosely tied together. Nitroglycerin is such an unstable substance that it needs only to be bumped a little in order to rearrange its atoms *quickly* (*very* quickly) into more stable, lower-energy combinations—a variety of gases. The energy released in the resulting explosion is the difference between the energy of the original nitroglycerin and the energy of the gases that it has rearranged its atoms into.

In general, if we can find a way to rearrange a substance's atoms into a lower-energy grouping, the "lost" energy must come out in some form, usually in the form of heat. When we burn coal or oil in air, we're giving their atoms (along with some oxygen atoms) an opportunity to rearrange themselves into lower-energy combinations—carbon dioxide and water. We can then collect the liberated energy as heat. The only reason we can't get energy out of water or stone is that we can't find any lower-energy arrangements of their atoms to help them transform themselves into. At least not without expending more energy than we'd get back.

In order to convert themselves into lower-energy atomic combinations, oil, natural gas, and gasoline—all of our usual fuels—must be offered oxygen to react with. Uranium atoms, however, don't need any such help. They can achieve a lower-energy state simply by splitting—dividing up their substance into two smaller atoms instead of one big one. The two smaller atoms happen to be tighter, more stable, lower-energy arrangements of subatomic particles than the original uranium atom was. The consequent decrease in energy is the energy of *nuclear fission*. Actually, it is only the *nucleus* of the uranium atom that does the splitting; the rest of the atom (the electrons) just goes along for the ride.

But all atoms are not capable of splitting their nuclei to give off energy. Only the very heaviest ones are susceptible to breaking apart like this. They are so heavy that they're actually a bit wobbly, and they will wobble themselves completely apart—split—at the slightest provocation. A nuclear reactor is essentially a very efficient provocateur. It tickles these wobbly uranium nuclei by lobbing *neutrons*, heavy, uncharged nuclear particles, at them, and that's all the encouragement they need to literally fall apart into more stable arrangements, giving off energy in the process.

NITPICKER'S CORNER:

Why is the uranium nucleus so unstable that it is eager to split in two?

All atomic nuclei are made up of particles called *nucleons*. A big nucleus like uranium's is a conglomeration of more than a couple of hundred of these particles, all crowded together into an incredibly tiny space. That's such a large number of objects to hold together that the nucleus's average grip on each one is rather weak. It's like trying to hold a bushel of golf balls together in your arms without benefit of a bushel basket.

The nucleus could improve its precarious situation and get a better grip on itself if it could split into two more easily manageable loads—two smaller (half-bushel) bunches of golf balls that could be held more tightly and securely. The two smaller bunches, being better controlled, would be less likely to break up. They would have less potential for unruly, energetic behavior—less of what scientists call potential energy.

But as Einstein taught us, energy is mass and mass is energy. So if the two smaller nuclei have less energy than the big nucleus had, they should show this by having less mass. The two "half-bushel" nuclei together do indeed weigh less than the single "bushel" nucleus, even though they contain the same number of "golf balls." If you add up the masses—the weights—of the two smaller nuclei that the uranium nucleus splits into, you'll find that they total about a tenth of a percent less than the mass of the original uranium nucleus. That tenth of a percent of "lost" mass shows up as a lot of energy, because according to good old $E=mc^2$ (see p. 227), a little bit of mass is equivalent to an enormous amount of energy.

A trifle hard to swallow, perhaps? But if these ideas weren't true, there wouldn't be such a thing as nuclear energy—or any of the hundreds of other nuclear goings-on that scientists see happening in their labs every day. Once we buy Einstein's proposition that energy and mass are interchangeable, we've bought into all of these happenings as perfectly natural consequences, and they shouldn't be the least bit surprising.

Well, maybe a wee bit.

Iron-ic Attraction

What makes a magnet attract iron? Why doesn't it attract aluminum or copper, for example?

Magnets are attracted only to other magnets. A piece of iron contains billions of tiny magnets, but copper and aluminum don't.

The only thing that the pole of a magnet will attract is the opposite pole of another magnet. It's exactly the same as with electric charges: The only thing that a positive electric charge will attract is a negative electric charge, and vice versa. With magnets, we call the two opposites "north" and "south" instead of "positive" and "negative." There is no direct force between an electric charge and something that isn't electrically charged. It's the same with magnets; no second magnet, no attraction.

(There are some crossover effects between electricity and magnetism; you can get magnetic attractions from moving charges and electric attractions from moving magnets. We'll consider stationary magnets only.)

Iron's atoms are tiny magnets because their negatively charged electrons—each iron atom has twenty-six of them—are spinning like tops as they circle the nucleus in the same way the Earth spins as it circles the sun. This spinning motion generates one of those "crossover" electric-magnetic situations, making their electric charges act like magnets. But most of iron's electrons are arranged in such a way that they're all matched up in pairs, and when spinning electrons pair up, they cancel each other's magnetism just as two bar magnets would: north pole to south pole and south pole to north pole.

Four of the iron atom's electrons, however, don't have partners, and because they're not paired up they impart a net, uncanceled magnetic effect to the atoms. Iron atoms, therefore, are magnetic and will be attracted to a magnet.

That's all very well, but iron isn't alone, by any means. Dozens of elements—even aluminum and copper—have unpaired electrons in their atoms and are therefore magnetic. Even oxygen atoms have unpaired electrons and are attracted to a magnet. You can't see it happen in the air, of course, but if you pour liquid oxygen onto a strong magnet in a lab, you'll see it stick.

This kind of magnetism that comes from unpaired electrons (Techspeak: *paramagnetism*) is quite weak, however. It is only about a millionth as strong as the kind of magnetism that we usually think of: iron being attracted to a magnet. But you can still observe it at home if you look closely.

TRY IT Place a carpenter's bubble level on the table and bring a strong magnet close to one end of the bubble. Observe closely, using one of the index marks on the tube as a reference point, and if the magnet is strong enough you'll see the air bubble move slightly toward the magnet. That's not because the oxygen in the air in the bubble is attracted to the magnet, however. You'd need a very powerful magnet to see that. It's because the liquid in the level is repelled from the magnet by a kind of paramagnetism. When the liquid moves one way, the bubble moves the other.

What is different about iron's much stronger kind of magnetism (Techspeak: *ferromagnetism*; *ferrum* is Latin for iron) is that in a piece of iron, the atomic magnets need not always be pointing in random, every-which-way directions, like a bunch of compasses in a magnet warehouse. If we stroke a piece of iron with a magnet, we can drag the iron's atoms into a lined-up arrangement, all pointing their north poles in the same direction and their south poles in the opposite direction.

Because of the precise sizes and shapes of the iron atoms, they will then remain in that lined-up arrangement without flopping back. This produces a very strong additive magnetic effect, millions of times stronger than the magnetism

of the individual atoms. The result is that the piece of iron has been *magnetized*: It has itself become a magnet and will attract other pieces of iron.

In only three elements, iron, cobalt, and nickel, are the sizes and shapes of the atoms exactly right for them to line up and stay that way. That's why these three metals are the only three ferromagnetic elements. Iron, however, is the strongest.

NO COMMENT NEEDED DEPARTMENT:

> The adult human body contains four to five grams of iron present in hemoglobin and myoglobin. [Actually, it's more like three.] Iron is very essential for our life and magnetism influences iron radically and magnificiently [sic]. . . . Magnets [therefore] have exceptional curative effects on certain complaints like toothache, stiffness of shoulders and other joints, pains and swellings, cervical sponcylitis [sic], eczema, asthma, as well as on chilblains, injuries, and wounds.
> (From *Magnetic Therapy*, a "health" tract distributed in a shopping mall to promote a clinic that "cures" with magnets.)

The World's Biggest BB Gun

If I dropped a BB from the top of the world's tallest building and it hit somebody on the head, would it kill him?

No. Pedestrians in the vicinity of Chicago's 1454-foot Sears Tower need not fear. Hatted or not, they are in little danger from purely scientific experiments such as yours. (We won't deign to discuss water balloons.)

What you undoubtedly have in mind is the *acceleration due to gravity*—the fact that a falling object will fall faster and faster as time goes by. That is indeed how the physics of falling operates. As an object falls, it is constantly being tugged upon by gravity, so no matter what its speed may be at any given instant, gravity is urging it onward to a still higher speed and it keeps going faster and faster. It accelerates. It is the same as if you were pushing a go-cart. So long as you keep pushing, the cart will keep going faster and faster. In a car, we would call this acceleration "pickup."

Might we not wonder, then, whether if we give our BB enough time to fall, it will eventually be traveling at bullet speed? Or why not the speed of light, for that matter? An actual calculation from the equations of gravity shows that after falling for a distance of 1454 feet, an object—any object—should be moving at a speed of 208 miles per hour. Beware below, indeed.

But wait. That's assuming that there is nothing at all between the mad bomber and his target. But there is. Air. And having to push its way through the air is bound to have a slowing-down effect on a falling object. We now have two opposing forces: the pull of gravity tending to speed the object up and the drag of air tending to slow it down.

Like any opposing forces in nature, these two clever forces work out a mathematically exact compromise. The slowdown from the air cancels out an equal portion of the speedup from gravity, thereby limiting the ultimate speed that the object will attain no matter how long it falls. It will fall faster and faster only up to a point, and from then on it will fall at a constant speed.

Of course, the air resistance will be different for the various objects that you might consider dropping off a building: much less resistance for a plucked chicken, for example, than for a feathered one. Therefore, the ultimate speed of different objects dropped through the air will be different. If there were no air, their speeds would all be the same after the same amount of time, no matter what their weights.

For a BB, the air resistance is such that its final speed at street level would be quite harmless, even to a bald head. And it would reach that final speed after falling only a few stories, so a scientific expedition to Chicago is entirely unnecessary.

Remember, of course, that it is not speed alone that determines the destructive power of a missile; it is *momentum*. Momentum is a combination of speed and weight. Even though a dropped bowling ball's final speed might not be bullet-like, its effects on a pedestrian could be rather severe because of its weight.

But you probably guessed that.

The Cosmic Boogie

They taught me in chemistry class that all atoms and molecules are in perpetual motion. But then they taught me in physics class that there's no such thing as perpetual motion, and that nothing can keep moving forever without being shoved. (Isaac Newton may not have put it quite that way.) So who's shoving all those atoms and molecules around?

Suppose that the Rockettes came on stage one at a time and did solos. You would agree that the intended effect would be lost, would you not? But that's exactly how school science curricula are designed: the chemistry and physics teachers do solo acts on separate stages, and there's no course in school called Putting It All Together.

Both of your recollections are correct, of course. The missing link is this: Nobody's pushing all those atoms and molecules around now, but they got one helluva shove some billions of years ago.

The movement of atoms and molecules, like the movement of anything else, is a form of energy called *kinetic energy* (from the Greek *kinema*, meaning motion). In the case of atoms and molecules, the kinetic energy is exhibited as an incessant flitting around and crashing into one another, restrained only by *bonds*, which is how chemists refer to the various kinds of attractions, or stickiness, between the particles. We call the collective motion of atoms and molecules *heat*.

The fact that all of these particles are in constant motion

doesn't mean that any visible-sized chunk of matter—a grain of salt, for example—is going to be bounding about like a Mexican jumping bean. The three billion billion atoms in that grain of salt (yes, I actually calculated it) are oscillating in all possible directions, so they cancel each other out. The salt grain isn't about to suddenly jump off your dining table. A beehive doesn't go galloping across the countryside just because the bees inside are flitting frantically about. (Actually, they're really pretty sedate inside the hive unless you rile them up.)

The big question, then, is where did all the particles in the world get their kinetic energy? Was there one great big, initial shove? Yes, indeed. All the matter in the universe obtained all of its energy at the moment of its creation in the "big bang" that, according to the most widely accepted theory, ignited the universe some ten or twenty billion years ago (cosmologists are still fighting over the exact date). And billions of years later, every particle in the universe is still shivering.

Not all at the same speed, however. When we add heat energy to a pot of soup on the stove, the soup particles will, *on the average*, be moving faster. And when we remove heat energy from a bottle of beer by putting it in the refrigerator, the beer particles will, *on the average*, move more slowly.

You know, of course, that what's going up on the stove and down in the fridge is the *temperature*: the average kinetic energy of the particles in a sample of matter, whether it is soup, beer, a human being, or a star. The key word is *average*.

Now we obviously can't climb into a pot of soup with a stopwatch and clock every one of its jillions of particles and average the speeds together to get its temperature. So we've had to invent a gadget called a thermometer. (It was invented by a man named Gabriel Fahrenheit.) The thermometer has a shiny, highly visible liquid in it—mercury—that expands up a glass tube when the temperature goes up and contracts down the tube when the temperature goes down. The mercury expands because of a chain reaction of collisions. The particles of the substance whose temperature we want to measure are colliding with the outer wall of the glass thermometer. This makes some of the glass particles collide with mercury particles inside the tube. The struck mercury particles are then moving faster than they were before, commanding more elbow room, and the mercury has to expand up the tube.

Thus, all the atoms and molecules in the universe are still shivering with primordial, universal energy, at different speeds depending on their temperature. And energy is all there is. It's the one and only currency in the universe. It can be converted from one form to another, just as money can be converted between the currencies of different nations. It can be lost by one body and gained by another, just as money can be transferred in a financial transaction. It can even be converted into mass, just as money can be converted into goods (see p. 226). The only things it can't do is be created (the mint went out of business right after the big bang) or be destroyed. We obtained a certain amount of it in the big bang, and we've been living on our budget ever since, in the form of heat and all the other forms that energy can be converted into.

In case you think that the sun is continually manufacturing new energy and sending it down to us as heat and light, think again. The sun and stars are just converting into these forms of energy some of the stash of energy that they already possess in the form of mass. Nothing new is being generated.

But won't the cosmic battery, charged up billions of years ago, ever die?

There is every reason to believe that it will. All the energy in the universe is gradually but relentlessly turning into something else: *entropy*, or disorder—total chaos (see p. 250). But don't worry too much about it. Long before that happens—only about six billion years from now as a matter of fact—the sun will have died.

BAR BET There is such a thing as perpetual motion, continuing as long as the universe lasts.

Mad About Metrics

Suddenly, soda pop and liquor come in liters, not quarts or fifths. Is this the first shot in a coming metric revolution? Do we really have to switch over to a whole new system of measurements? What's wrong with the system we have now?

Among all the nations of the world, only four great powers—Brunei, Myanmar (Burma), Yemen, and the United States of America—have not yet adopted the metric system of measurement. Is it possible that the rest of the world is onto something that has thus far eluded these four?

Let's see how our creaky and quirky English system of measurement (which even the English don't use anymore) might be improved. Here is the list of ingredients in a recipe for coffee cake:

1⅓ cups sour cream
1¼ teaspoons baking soda
1¾ teaspoons baking powder
1¾ cups cake flour
2 eggs
1½ cups sugar
½ cup butter

Now suppose that you want to make half the recipe. Your assignment: Cut the ingredient amounts in half.

Let's see, now. Half of one and one-third is, er . . . Well, half of one and one-fourth is . . . um . . . Half of one and three-fourths . . . Well, there are eight ounces in a cup (or is it sixteen?), so half of one and three-fourths cups of flour is one and three-fourths times eight divided by two, or . . . Why don't I just take half of two eggs—I can do that in my head—and guess at the rest of the stuff?

Good luck.

Now let's imagine a brave new world in which everything is in metric units. The coffee cake recipe would work like this:

Full recipe	*Half recipe*
320 grams sour cream	160 grams sour cream
6 grams baking soda	3 grams baking soda
9 grams baking powder	4½ grams baking powder
230 grams cake flour	115 grams cake flour
2 eggs	1 egg
300 grams sugar	150 grams sugar
110 grams butter	55 grams butter

Simplicity itself, no?

Now all you have to know is what in the world a gram is, right? Not really. If in the brave new world you own a gizmo that measures everything out in grams, what do you care how big a gram is? Just measure out 160 of them, 3 of them, 4 ½ of them and so on, *whatever* they are. Do you really know what an "ounce" is? All you know is that it's a certain amount of stuff that some person or persons unknown, for reasons unknown, decided upon a long, long time ago.

Moreover, we must constantly wrestle with three kinds of ounces: fluid, avoirdupois, and troy; and they're all different. They don't even measure the same thing; two of them measure weight and one measures volume.

A gram is a unit of weight. Weighing things out on a scale is a lot more accurate and reproducible than filling up measuring cups, teaspoons, and tablespoons, especially with messy stuff like butter. Okay, so you'll have to buy a kitchen scale. Serious chefs already weigh out their ingredients.

Now, out of the kitchen and into the workshop. You have a board that measures seven feet, nine and five-eighths inches, and you need to cut it into three equal lengths. Again, good luck with the calculation. (The answer, which you can arrive at in substantially less than an hour, is two feet, seven and seven thirty-seconds inches, more or less.) In the brave new world, you would measure the board with a meter stick and find that it is 238 centimeters long. One-third of that is 79.3 centimeters. End of problem.

Note that you didn't have to know or care that there are 2.54 centimeters in an inch, any more than you had to know that there are 28.35 grams in an ounce when you weighed out your cake ingredients. Just think of a centimeter as the distance between two adjacent numbers on the stick and a gram as one of those little divisions on the scale.

Many people despair of ever learning the metric system because the units—the grams and centimeters and so on—are hard to visualize in terms of familiar ounces and inches. In other words, it's the *conversion* between the old and new systems that is troublesome. And indeed it is. Who wants to keep messing around with 2.54s and 28.35s all the time? There's no doubt that it is going to be terribly awkward to convert everything in the United States—from recipes to road maps, not to mention all of our industrial production facilities—to the metric system. Nobody argues with that.

But that's the wrong reason for resisting the metric system. Don't we now have to perform ridiculously difficult conversions every day in the English system? Twelve inches in a foot; 3 feet in a yard; 1760 yards in a mile; 16 avoirdupois ounces in a pound, 16 fluid ounces in a pint; 2 pints in a quart, 4 quarts in a gallon, and so on. Not to mention wrestling with pecks, bushels, barrels, fathoms, knots, and literally hundreds of other crazy units.

In the metric system, there is only one unit for each type of measurement. And the only conversion numbers you'll need are 10, 100, and 1000; not 3, 4, 12, 16, or 5,280. There are 100 centimeters in a meter, 1000 meters in a kilometer, 1000 grams in a kilogram, and so on. Using metrics is simplicity itself, as evidenced by the fact that every schoolchild and housewife throughout 94 percent of the world's population has no trouble at all with it.

Once our awkward transition period is over, life will be beautiful. But the longer we wait, the tougher the transition is going to be.

The United States of America has a real bad toothache and is procrastinating about seeing the dentist.

You didn't ask, but . . .

Some weather reports are already giving temperatures in degrees Celsius, but the conversion formulas they gave us in school are complicated and impossible to remember. Is there an easy way to convert Celsius to Fahrenheit?

Yes, there is a much simpler way, and it's a shame they don't teach it in school. Once those complicated formulas with all their parentheses and 32s got into a textbook somewhere, they seem to have taken on a life of their own.

Here's the simple method:

> *To convert a Celsius temperature to Fahrenheit, just add 40, multiply by 1.8, and subtract 40.*

That's all there is to it.

For example, to convert 100 degrees Celsius, we'll add 40 to get 140, multiply by 1.8 to get 252, and then subtract 40 to get 212. What do you know! That's the boiling temperature of water: 100 degrees Celsius equals 212 degrees Fahrenheit.

The great thing about this method is that it works in both directions, to wit:

> *To convert a Fahrenheit temperature to Celsius, just add 40, divide by 1.8, and subtract 40.*

Example: to convert 32 degrees Fahrenheit to Celsius, we'll add 40 to get 72, divide by 1.8 to get 40, then subtract 40 to get—and there you are!—0. That's the freezing temperature of water: 32 degrees Fahrenheit equals 0 degrees Celsius.

All you have to remember is whether to multiply or divide by 1.8. Hint: Fahrenheit temperatures are always bigger numbers than Celsius. So when you're going toward Fahrenheit, you multiply.

NITPICKER'S CORNER:

Why does this method work? Because of the ways that Messrs. Fahrenheit and Celsius set up their temperature scales. It turns out accidentally that 40 degrees below zero on either scale represents exactly the same temperature. So adding 40 puts them on the same basis, so to speak. Then all we have to do is correct for the different sizes of the degrees (a Celsius degree is exactly 1.8 times as big as a Fahrenheit degree), and finally we remove the artificial 40 that we added.

A more detailed proof of why this method works would constitute a smaller nit than we care to pick at the moment. But it always works, and it is exact.

Weighty Matters

Why is helium lighter than air? For that matter, why is anything lighter or heavier than anything else?

Everything is made of particles: atoms and molecules. But it's not simply that some particles are lighter than others, although that's a big part of it. It's also that some particles are packed more tightly together than others.

Lead is denser than—i.e., heavier than the same volume of—water, mostly because lead atoms are more than eleven times heavier than water molecules. But even if the lead and water particles were the same weight, there could be a difference in density because of the packing. For example, liquid water is denser than solid water (ice), even though they're both made of the same particles, namely water molecules. But in the liquid, the molecules are packed together more tightly than they are in the solid. So when somebody claims that one substance is denser than another because its particles are heavier, they're not necessarily telling the whole story.

Gases are a whole different ball game from liquids and solids, though, because they don't pack together at all; their molecules are flying around in space completely free of one another. At the same pressure, all gas molecules will be packed together (actually, *un*packed) to exactly the same extent—that is, they'll be separated by the same average distance, whether they're helium atoms or air molecules.

Thus, packing has nothing at all to do with which gas is denser. Helium is about one-seventh as dense as air at the same pressure, simply because its particles weigh one-seventh as much as the average air particle.

Just what you thought, right? But perhaps for the wrong reasons.

Hey! Whose DNA Is This?

What are all those little black smudges that the experts keep waving around in courtrooms as "DNA evidence" of this or that? Are they the DNA itself?

No. Those ladders of fuzzy black dashes are just a way of making visible to jurors and other ardent scholars of biochemical science certain things that are too small to see, even with a microscope. They're the end result of a number of laboratory manipulations that never get explained in the courtroom. But before we describe them, will the *real* DNA please stand up?

DNA is the most intricate and awe-inspiring substance on Earth, but it is not too hard to understand if we stay away from the big words and stop just this side of more-than-you-want-to-know.

Suppose that you are Mother Nature, and you want to set up a general scheme of life that will work for all living things, both plant and animal. The biggest problem you face is how to get from one generation to another. After all, manufacturing one exquisite rose, cockroach, or horse, no matter how difficult that may be, isn't going to get you very far unless you give it the power to make more roses, cockroaches, and horses. How, then, can a rose beget a rose? How can a horse inform its offspring that it should be a horse, rather than a blade of grass or a cockroach—having four legs instead of six, no chlorophyll or antennae, and so on, and on and on?

There are an enormous number of explicit specifications that must be noted and carried out to ensure that each succeeding generation follows the same pattern. How has Mother Nature arranged to record and play back, time after time, without benefit of pencil and paper, videotape, or CD ROM, the immense amount of complex information that, taken all together, says "horse"?

Answer: She writes it all down on strips of a remarkable substance called DNA, as if on strips of recording tape.

DNA is a merciful abbreviation for *deoxyribonucleic acid*. This substance is made up of certain specific clusters of atoms, lined up into long ribbons that are twisted into spirals and then coiled up into compact little packages and tucked into the nuclei of virtually every cell of every part of every life-form on Earth, from six-ton elephants to one-celled bacteria and lawyers.

The information on the DNA ribbons is written in a code. The code consists of the exact sequences in which those clusters of atoms are arranged along the ribbon. If you think of the atom clusters as words, their sequences are sentences. Specific sequences of atom clusters convey specific pieces of information, just as specific sequences of words do in a sentence.

Scientists refer to the atom-cluster "words" as *nucleotides* and the "sentences" as *genes*. Each gene "sentence" states an essential bit of information about what the baby horse—or cockroach or human being—shall or shall not be. Genes even distinguish each individual baby from all others. In a single human DNA ribbon, there are so many "words" (a few million, perhaps), combined into so many gene "sentences" (maybe a hundred thousand), that except for identical twins, no two individuals among the five billion people now on Earth—or among all those who have gone before—should have exactly the same combination.

Just imagine the odds. If you had a basket with a few million words in it and you reached in blindfolded and picked out enough words, one by one, to make a (rather long) book of a hundred thousand sentences, what do you think your chances are of repeating the process and getting exactly the same collection of words in the same sequence—that is, of getting exactly the same book? In the case of humans, the odds are even more extreme because of historical and geographical isolation: the odds of exactly duplicating a certain black African baby in a Swedish maternity ward are even slimmer than simple mathematics would indicate.

Aha! Then if every human being on Earth has a unique set of genes on his or her DNA ribbons, can we tell what characteristics an individual has by examining his or her DNA? In principle, yes, except that we haven't yet worked out the entire sequence of genes on anyone's DNA. But if DNA is found in every cell in the body, from skin to blood, hair, fingernails, and semen, couldn't we identify the perpetrator of a crime, for example, by matching a suspect's DNA with the DNA from cells found at the scene? Definitely. And that's what forensic DNA analysis is all about.

How do they do it? They extract the DNA from the cell samples and treat it with enzymes that "grow" the DNA—make repeated identical copies of it—until there is enough to work with. Other enzymes then cut up the ribbons into various manageable-sized fragments, like cutting up a book into various pages, paragraphs, sentences, and phrases. Then the technicians spread out all the cut-up fragments according to their sizes (I'll tell you how) and compare exactly which arrangements of words show up in both of the samples that they want to compare. Same fragments means same DNA and same person.

Think about it. If you can cut two books into hundreds of pieces and wind up with even half a dozen identical pages or assortments of paragraphs in exactly the same order, then by golly you've got two copies of the same book. (Or one heck of a case of plagiarism.)

Now, about the infamous black smudges. Those ladders of thick, black lines are made by the fragments of DNA, which have been spread out along a kind of racetrack according to their sizes in an electrical apparatus. Technicians give the fragments a negative electric charge and allow them to drift slowly along a surface toward a positive electric pole. The smallest, lightest fragments drift fastest and travel farthest, winding up at the top of the ladder when the race ends; heavier fragments lag behind to various degrees. Thus, they are spread out according to their sizes.

The invisibly small groups of separated DNA fragments are made radioactive, so that their radiations will expose spots on a sheet of photographic film, thus visually revealing their final locations on the racetrack. That developed sheet of film, bearing black exposure marks wherever the fragments wound up at the end of the race, is what the scientists compare, thereby comparing the DNA structures of the two samples. The same finish positions at the end of the race indicate the same DNA and therefore the same individual, with odds that can be as high as hundreds of trillions to one.

Of course, there is always a slim chance that the murderer was a horse.

Use It and Lose It

To save energy and resources, we're recycling all sorts of things these days. Can we recycle energy itself?

Absolutely, if by recycling you mean transforming something into a more useful form. We do it all the time. Power plants transform water, coal, or nuclear energy into electricity. In our kitchen toasters we transform electrical energy into heat energy. In our automobile engines we transform chemical energy into motion (kinetic energy). The different forms of energy are all interchangeable; all we have to do is invent an appropriate machine to do the job.

But there's a catch—perhaps the biggest catch in the entire universe: Every time we convert energy, we lose some of its value. That's not just because our gadgets are inefficient or because we're sloppy; it's more fundamental than that. It's like converting currency in a foreign country; there is a cosmic exchange agent who inevitably takes a little cut out of each transaction. The name of this cosmic exchange agent is the Second Law of Thermodynamics.

It's really a good-news-bad-news joke.

First, the good news. That's the Law of Conservation of Energy, also known as the First Law of Thermodynamics. It says that energy cannot be created or destroyed. It can be changed back and forth from one of its many forms to another—heat, light, chemical, electrical, mass, and so on—but according to the First Law the *amount* of it must always stay the same; energy never just disappears. The amount of mass-energy in the universe was fixed at the time of its creation (see p. 236). We can never run out of energy.

Great! Then all we have to do is keep converting and reconverting our energy into whichever form we happen to need at the moment—light from a bulb, electricity from a battery, motion from an engine—and keep using it over and over. We'll recycle energy just as we recycle aluminum cans, right?

Unfortunately, wrong. Here's the bad news. The Second Law of Thermodynamics says that every time we convert energy from one form to another, we lose a little of its usefulness. We can't lose any energy itself—the First Law prohibits that—but we lose some of its *ability to do work*. And if you can't do work with it, what good is energy, anyway?

The reason that some work strength is lost is that every time we convert energy from one form into another, some of it winds up as heat energy whether we want it to or not.

About 60 percent of the energy in the coal that they burn down at the local power plant is wasted as heat; only about 40 percent of it winds up as electricity, and much of that is lost on its way to you through those overhead wires. Then, 98 percent of the electrical energy you put into a light bulb is wasted as heat. Much of the chemical energy in gasoline goes out the radiator and tailpipe of your car as heat.

Even if all these complex operations were 100 percent efficient, some heat would inevitably be lost. Even when water turns a waterwheel, a little bit of the water's energy winds up as frictional heat in the wheel's bearings.

Expecting no heat to be formed at all is like expecting no friction. And expecting no friction would be expecting a machine to keep running forever without slowing down. Perpetual motion. Energy from nowhere. And that's impossible. (See the First Law.) Therefore, wherever energy is being put to work, some heat must be formed.

But heat is still energy, isn't it? Sure it is. Then why can't we just take that heat and put it back to work as usable energy?

Here's the *real* bad news of the Second Law: We can indeed do that, but not completely. While other forms of energy can be converted 100 percent into heat, heat cannot be converted 100 percent into any other form. Why? Because heat is a random, disordered motion of molecules (see p. 236). And once your energy is in that chaotic condition, you just can't get a full complement of useful work out of it. Just try to plow a field with a "team" of horses that are running around in all directions.

So little by little, as the world spins on, all forms of energy are relentlessly being converted into irretrievable heat. The world's energy is gradually turning into a useless, chaotic motion of particles. The more we use, the more we lose.

The universe is running down like a cheap battery.

We're on a one-way street, headed down.

Have a nice day.

Why Does That Happen, Daddy?

This may be a dumb question, but what makes things happen or not happen? Water will flow downhill, but not up. I can put sugar in my coffee, but if I put in too much I can't get it out again. I can burn a match, but I can't unburn it. Is there some cosmic rule that determines what can happen and what can't?

There's no such thing as a dumb question. Actually, yours is perhaps the most profound question in all of science. Nevertheless, it does have a fairly simple answer—ever since a genius by the name of Josiah Willard Gibbs figured it all out in the late nineteenth century.

The answer is that everywhere in nature there is a balance between two fundamental qualities: *energy*, which you probably know something about; and *entropy*, which you probably don't (but soon will). It is this balance alone that determines whether or not something can happen.

Certain things can happen all by themselves, but they can't happen in the opposite direction unless they get some outside help. For example, we could make water go uphill by hauling it or pumping it up. And if we really wanted to, we could get that sugar back out of the coffee by evaporating the water and then chemically separating the sugar from the coffee solids. Unburning a match is quite a bit tougher, but given enough time and equipment, a small army of chemists could probably reconstruct the match out of all the ash, smoke, and gases.

The point is that in each of these cases a good deal of meddling—energy input from outside—is required. Left entirely to herself, Mother Nature allows many things to happen spontaneously, all by themselves. But others will *never* happen spontaneously, even if we wait, hands-off, until doomsday. Nature's grand bottom line is that if the balance between energy and entropy is proper, it will happen; if it is not, it won't.

Let's take energy first. Then we'll explain entropy.

In general, everything will try to decrease its energy if it can. At a waterfall, the water gets rid of its pent-up gravitational energy by falling down into a pool. (We can make that cast-off energy turn a water wheel for us on the way down.) But once the water gets down to the pool, it is "energy-dead," at least gravitationally speaking. It can't get back up to the top. A lot of chemical reactions will happen for a similar reason: The chemicals are getting rid of their pent-up energy by spontaneously transforming themselves into different chemicals that have less energy. The burning match is one example.

Thus, other things being equal, nature's inclination is that *everything will lower its energy if it can*. That's rule number one.

But *decreasing energy* is only half the story of what makes things happen. The other half is *increasing entropy*. Entropy is just a fancy word for *disorder*, or randomness; the chaotic, irregular arrangement of things. At the scrimmage, football players are lined up in an orderly arrangement—they are not disorderly, and they therefore have low entropy. After the play, however, they may be scattered all over the field in a more disorderly, higher-entropy arrangement.

It's the same for the individual particles that make up all substances: the atoms and molecules. At any given time, they can be in an orderly arrangement, in a highly disordered jumble, or in any kind of arrangement in between. That is, they can have various amounts of entropy, from low to high.

But other things (namely, energy) being equal, Nature's inclination is that everything tends to become more and more disorderly—that is, *everything will increase its entropy if it can*. That's rule number two. There can be an "unnatural" increase in energy as long as there is a more-than-compensating increase in entropy. Or, there can be an "unnatural" decrease in entropy as long as there is a more-than-compensating decrease in energy. Got it?

So the question of whether or not a happening can occur in nature spontaneously—without any interference from outside—is a question of balance between the energy and entropy rules.

The waterfall? That happens because there is a big energy decrease; there's virtually no entropy difference between the conditions of the water at the top and at the bottom. It's an energy-driven process.

The sugar in the coffee? It dissolves because there's a big entropy increase; sugar molecules swimming around in coffee are much more disorderly than when they were tied neatly together in the sugar crystals. Meanwhile, there is virtually no energy difference between the solid sugar and the dissolved sugar. (The coffee doesn't get hotter or colder when the sugar dissolves, does it?) It's an entropy-driven process.

The burning match? Obviously, there's a big energy decrease; the pent-up chemical energy is released as heat and light. But there is also a huge entropy increase; the billowing smoke and gases are much more disorderly than the compact little match head was. So this reaction is doubly blessed by nature's rules, and it occurs with great gusto the instant you provide the initiating scratch. It's driven by both energy and entropy.

What if we have a process in which one of the quantities,

energy or entropy, goes the "wrong way"? Well, the process can still occur if the other quantity is going the "right" way strongly enough to overcome it. That is, energy can increase as long as there's a big enough entropy increase to counterbalance it; and entropy can decrease as long as there's a big enough energy decrease to counterbalance it.

What J. Willard Gibbs did was to devise and write down an equation for this energy-entropy balance. It happens that if this equation comes out with a negative sign, the process in question is one that Mother Nature allows to occur spontaneously. If it comes out with a positive sign, the process is impossible. Absolutely impossible, unless human beings or something else sidesteps the rules by bringing in some energy from outside.

By using enough energy, we can always overpower nature's entropy rule that everything tends toward disorderliness. For example, with enough effort we could collect, atom by atom, the ten million tons of dissolved gold that are distributed throughout the Earth's oceans, sitting there just for the taking. But it is dispersed through 324 million cubic miles (1.35 billion cubic kilometers) of ocean in a random, incredibly high-entropy arrangement. The problem is that the energy necessary to separate and purify it would cost a lot more than the value of the gold.

In a fit of fervor over the laws of mechanics, Archimedes (287–212 B.C.) is reputed to have said, "Give me a lever long enough and a place to stand, and I will move the world." If he had known about entropy and apple pie, he might have added, "Give me enough energy and I will reassemble the world into apple-pie order."

Buzzwords

Acceleration: Moving faster and faster as time goes by.

Acid, base, and salt: Acids and bases are opposite kinds of chemicals that neutralize one another, forming water and a salt. Table salt is the most common salt. Common acids are carbon dioxide and vinegar. Common bases are ammonia and lye.

Alloy: A metal that has been made by melting together two or more pure metals.

Atom: A very tiny particle, of which all substances are made. There are 110 known kinds of atoms. Atoms are almost always joined together in various combinations to form *molecules*.

Calorie: An amount of heat energy. As chemists use the word, a calorie is the amount of heat that it takes to raise the temperature of one gram of water by one degree Celsius. The food-energy *Calorie* used by nutritionists is equal to one thousand of these chemists' calories. To avoid confusion in this book, we use the lower-case *calorie* for the chemist's and the upper-case *Calorie* for the nutritionist's.

Capillary: A very thin tube, or any very thin space through which a liquid can flow. Water and some other liquids creep automatically into such thin spaces because their molecules are attracted to the walls of the tube.

Carbohydrates: A family of plant chemicals that includes starches, sugars, and cellulose.

Centrifugal force: The force that makes things tend to fly outward when you swing them in a circle. If you have ever heard of *centripetal force*, forget it; it's unnecessarily confusing.

Chemical compound: A pure, definable substance whose molecules are made up of definite types and numbers of atoms. We don't often see pure chemical compounds in nature because almost everything in the world is made up of combinations and mixtures of them.

Condensation: When a vapor cools enough to become a liquid, it is said to condense. Condensation is the reverse of boiling, in which a liquid gets hot enough to turn into a vapor.

Cosine: What you get by pressing the *cosine* button on your calculator.

Crystal: A solid that is made up of a regular geometric arrangement of particles. The solid reflects this regular internal arrangement by having a regular geometric outer shape.

Density: A measure of how heavy a given volume (bulk) of a substance is. A cubic foot of water, for example, weighs 62.4 pounds. The density of water is therefore 62.4 pounds per cubic foot. In metric units, the density of water is one gram per cubic centimeter. For comparison, the density of lead is 11 grams per cubic centimeter.

Dissolving: When a substance dissolves in water, it seems to disappear because it actually does come apart; its molecules separate from one another and mix in intimately amongst the molecules of water. This mixture is called a *solution*. The "strength" of a solution is an indication of how much of the substance is dissolved in any given amount of water. For reasons of their own, chemists insist on using the word *concentration* instead of *strength*.

Electromagnetic radiation: Pure energy in wave form, traveling through space at the speed of light. Known types of electromagnetic energy range from radio waves to microwaves to light (both visible and invisible) to X rays to gamma rays. Electromagnetic waves have a *wavelength* and a *frequency* of vibration; the shorter the wavelength, the higher the frequency and energy.

Electron: A tiny, negatively charged particle. Its native habitat is outside the *nucleus*, the extremely heavy core, of an atom. Electrons are easily detached from their atoms and can move about on their own.

Enzyme: A natural *catalyst*—a substance that speeds up a chemical process without itself being used up or changed in any way. Enzymes in plants and animals make otherwise too-slow life processes take place at a reasonable speed.

Heat: A form of energy, manifested by the motion of the atoms and molecules in a substance.

Heat capacity: The amount of added heat that is required to raise the temperature of a substance by a certain number of degrees. Water, for example, soaks up a great deal of heat before getting much hotter; it has a high heat capacity.

Heat of fusion: The amount of heat it takes to melt a solid, usually expressed as the number of calories needed to melt one gram of it.

Inorganic or organic compound: Chemists have divided all chemicals into two classes: inorganic and organic. Organic compounds contain carbon atoms in their molecules; inorganic compounds don't. Almost all of the chemicals involved in plant and animal life are organic.

Ion: An ion is an atom or group of atoms that have an electric charge, obtained by losing some of their electrons or by gaining extra ones. Most minerals exist as ions, rather than as uncharged atoms or molecules.

Kinetic energy: Kinetic energy is energy that is in the form of action, or motion. A pitched baseball has obvious kinetic energy. But heat is also a form of kinetic energy because it consists of the movement of the atoms and molecules in an object, even though the object itself may not be moving.

Molecule: A tiny particle, of which almost all substances are made. Different substances are different because their molecules are different in composition, arrangement, size, or shape. Molecules, in turn, are made of even tinier particles called *atoms*. Atoms, in their turn, are made of *electrons*, which are distributed around a *nucleus*.

Nucleus (plural, nuclei): The heavy central core of the atom, containing virtually all of the atom's mass or weight. It is thousands of times as heavy as the atom's electrons.

Polar: A polar substance is made up of molecules whose electrons are more concentrated at one end than at the other, which makes that end of the molecule negatively charged, compared with the other end. Such a molecule responds to electric and magnetic forces, whereas a nonpolar molecule would be unaffected. Water molecules are strongly polar, which gives water some unique properties.

Polymer: A substance whose large molecules are made up of many smaller molecules, all tied together. Plastics and proteins are polymers.

Potential energy: Energy that is somehow stored up and waiting to be released to do useful work. Examples: gravitational potential energy (a boulder poised on the rim of a canyon), chemical potential energy (a stick of dynamite), and nuclear potential energy (a bunch of uranium atoms).

Pressure: The amount of force that is being applied to each and every unit of an area. Often measured in pounds per square inch.

Protein: A *polymer* found in plants and animals, whose large molecules have been formed by amino acid molecules condensing together. Amino acids are nitrogen-containing, organic chemical compounds that are essential to human metabolism.

Redox reaction: A chemical reaction in which electrons are being passed from one kind of atom, molecule, or ion to another.

Spectrum: A display of all the wavelengths of radiation that a given substance emits or absorbs. The sun emits a broad spectrum of radiation, including the *visible spectrum*: the rainbow of colors that the human eye can see.

Temperature: A number that expresses the average *kinetic energy*, or motion energy, of all the particles in a substance.

Index

About the Author

Robert L. Wolke (Ph.D., Cornell University) is a scientist turned journalist. Currently professor emeritus of chemistry at the University of Pittsburgh, he has long been known for his ability to make science both understandable and enjoyable to liberal arts students. He has also taught at the Universities of Florida, Puerto Rico, and Oriente in Venezuela, and has carried out research in nuclear chemistry at Oak Ridge National Laboratory and at the University of Chicago's Enrico Fermi Institute. In addition to numerous research papers, he is the author of the books *Impact: Science on Society* and *Chemistry Explained*, and is a frequent contributor of articles and essays to magazines and newspapers across the country. He lives, writes, and cooks in Pittsburgh with his wife, Marlene Parrish, a food writer and restaurant consultant.